Make an Estimate

Round the numbers in the problems and add mentally to estimate the sum. (Round to the place that makes sense for each number or problem.)

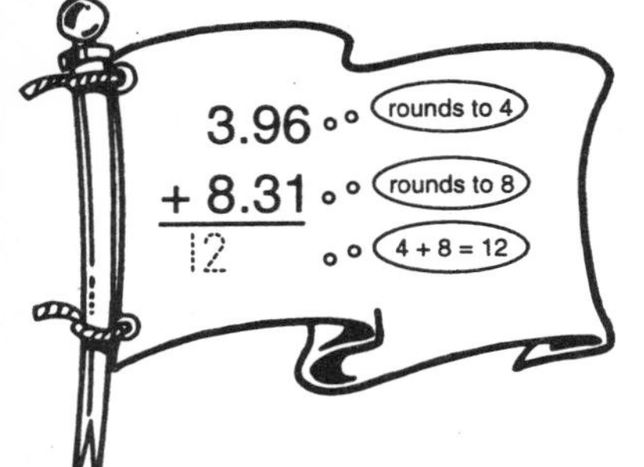

A.

2.04	6.014	10.33	3.49
+ 5.75	+ 3.275	+ 17.4	+ 15.52

B.

87.43	1.28	61.29	22.14	2.4
+ 80.40	+ 9.40	+ 59.24	+ 95.52	+ 39.6

C.

5.37	29.34	70.16	256.11	49.99
+ 3.79	+ 34.56	+ 45.42	+ 29.93	+ 34.56

D.

24.92	74.89	19.55	124.95	9.95
76.38	27.36	4.46	59.50	41.45
+ 67.21	+ 56.03	+ 20.26	+ 100.40	+ 60.88

E.

21.07	70.88	26.97	13.06	434.15
39.54	12.45	39.51	26.91	202.01
+ 60.24	+ 3.98	+ 59.94	+ 5.27	+ 46.82

Name ______________________________

Rough Estimates

Round the numbers in the problems and subtract mentally to estimate the subtraction. (Round to the place that makes sense for each number or problem.)

Example:

57.06 (rounds to 60)
− 35.98 (rounds to 40)
20 (60 − 40 = 20)

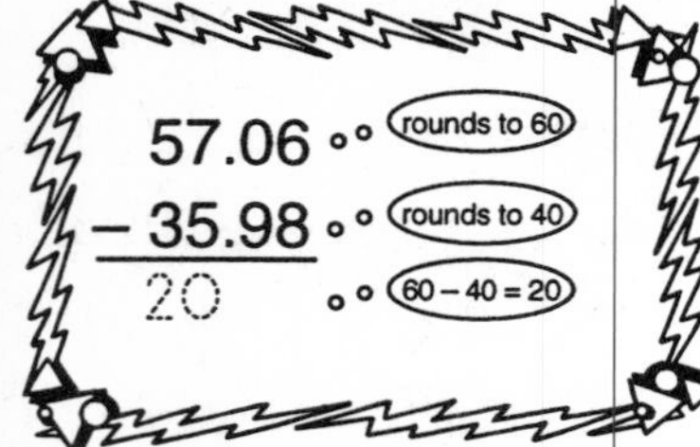

A.	9.67 − 3.81	5.03 − 4.21	11.20 − 9.67	7.83 − 1.96	
B.	12.40 − 10.66	19.05 − 4.41	32.79 − 12.19	12.89 − 5.24	9.36 − 5.44
C.	69.05 − 19.55	10.71 − 10.69	52.74 − 13.14	46.83 − 20.91	84.37 − 79.66
D.	55.70 − 51.42	68.30 − 45.75	8.13 − 7.95	32.78 − 18.29	7.83 − 6.10
E.	42.90 − 19.72	18.05 − 8.68	9.46 − 1.78	32.13 − 1.97	42.73 − 10.99
F.	237.94 − 99.75	481.3 − 76.1	47.85 − 19.99	9.65 − 5.84	10.039 − 9.876

Name ______________________________

Sum It Up

Find the sums.

A.	6.2 + 1.73 7.93	0.525 + 0.139	12.36 + 8.75	7.65 + 3.34	0.68 + 0.4
B.	72.88 + 14.75	9.89 + 42.69	0.25 + 0.73	5.9 + 6.057	38.93 + 17.6
C.	4.769 + 3.825	78.9 + 32.6	34.9 + 67.85	1.79 + 3.826	36.29 + 28.75
D.	4.7 8.8 + 0.45	439.6 7.049 + 12.32	72.6 123.12 + 2.4	2.36 3.78 + 2.67	42.83 75.6 + 36.356
E.	7.84 65.3 + 238.72	179.6 4.98 + 56.43	2.368 3.26 + 0.471	27.6 3.84 + 51.09	

Name ______________________________ Subtracting decimals

Decimal Differences

Find the differences.

A.	265.3 − 121.44 143.86	3.74 − 1.88	52.67 − 24.7	57.19 − 19.88	
B.	5.25 − 3.87	0.85 − 0.68	51.04 − 22.63	70.00 − 16.95	143.79 − 88.81
C.	26.85 − 15.97	71.35 − 4.661	73.21 − 56.56	54.135 − 27.950	95.41 − 8.72
D.	23.9 − 18.72	44.04 − 28.15	1.343 − 0.975	680.3 − 136.9	7.342 − 2.617
E.	56.53 − 17.6	213.06 − 4.8	1.04 − 0.999	8.35 − 2.967	4.0 − 2.91
F.	439.3 − 97.42	83.81 − 7.96	100.1 − 83.79	17.31 − 14.9	6.01 − 3.07

Name ____________________

Product Estimates

Estimate the products by rounding the numbers and multiplying. (Round to the place that makes sense for each number or problem.)

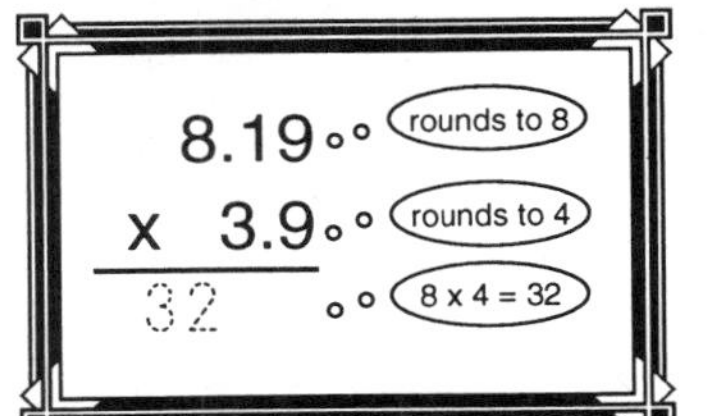

A.	32.9 x 8.6	43.82 x 17.9	8.97 x 56.3	27.87 x 3.56	
B.	52.7 x 1.87	19.97 x 4.95	3.7 x 8.3	6.47 x 7.19	8.96 x 4.1
C.	71.974 x 8.6	67.46 x 39.28	319.3 x 86.76	909.36 x 6.2	3.826 x 7.4
D.	9.15 x 57.8	6.45 x 8.59	7.83 x 0.84	48.42 x 2.26	8.99 x 2.3
E.	3.62 x 7.8	26.9 x 7.3	43.46 x 37.3	808.2 x 5.8	29.5 x 0.137
F.	63.40 x 4.91	29.2 x 8.71	73.9 x 2.03	57.8 x 6.34	98.4 x 0.5

Name ______________________________

Multiplying Decimals

Find the products.

A.	$\begin{array}{r} 3.5 \\ \times\ 6.7 \\ \hline \end{array}$	$\begin{array}{r} 8.09 \\ \times\ 5.7 \\ \hline \end{array}$	$\begin{array}{r} 12.5 \\ \times\ 0.74 \\ \hline \end{array}$	$\begin{array}{r} 9.4 \\ \times\ 2.7 \\ \hline \end{array}$	$\begin{array}{r} 12.8 \\ \times\ 3.5 \\ \hline \end{array}$
B.	$\begin{array}{r} 5.12 \\ \times\ 7.6 \\ \hline \end{array}$	$\begin{array}{r} 9.12 \\ \times\ 6.8 \\ \hline \end{array}$	$\begin{array}{r} 0.73 \\ \times\ 4.2 \\ \hline \end{array}$	$\begin{array}{r} 5.6 \\ \times\ 8.3 \\ \hline \end{array}$	$\begin{array}{r} 6.9 \\ \times\ 5.4 \\ \hline \end{array}$
C.	$\begin{array}{r} 8.42 \\ \times\ 7.3 \\ \hline \end{array}$	$\begin{array}{r} 7.58 \\ \times\ 4.8 \\ \hline \end{array}$	$\begin{array}{r} 53.7 \\ \times\ 6.9 \\ \hline \end{array}$	$\begin{array}{r} 4.86 \\ \times\ 3.7 \\ \hline \end{array}$	$\begin{array}{r} 6.45 \\ \times\ 7.6 \\ \hline \end{array}$
D.	$\begin{array}{r} 7.25 \\ \times\ 1.89 \\ \hline \end{array}$	$\begin{array}{r} 5.62 \\ \times\ 3.84 \\ \hline \end{array}$	$\begin{array}{r} 3.79 \\ \times\ 1.01 \\ \hline \end{array}$	$\begin{array}{r} 1.23 \\ \times\ 3.7 \\ \hline \end{array}$	

Use estimation to see if your answers are reasonable.

Name ____________________

Quotient Estimates

Estimate the quotients by rounding the numbers and dividing. (Round to the place that makes sense for each number or problem.)

176.2 ÷ 6.3
That's close to 180 ÷ 6, so the quotient will be about 30.

A. 176.2 ÷ 63 = ____________

B. 301.38 ÷ 5.3 = ____________

C. 11.93 ÷ 3.2 = ____________

D. 12.398 ÷ 4.1 = ____________

E. 482.04 ÷ 8.32 = ____________

F. 359.6 ÷ 8.5 = ____________

G. 64.26 ÷ 8.3 = ____________

H. 354.64 ÷ 69.3 = ____________

I. 413.66 ÷ 5.1 = ____________

J. 3,112.4 ÷ 5.2 = ____________

K. 409.4 ÷ 48.4 = ____________

L. 54.43 ÷ 9.32 = ____________

M. 48.14 ÷ 6.4 = ____________

N. 544.82 ÷ 9.1 = ____________

162.58 ÷ 4.4 = ____________

14.948 ÷ 5.1 = ____________

266.7 ÷ 3.1 = ____________

12.345 ÷ 3.3 = ____________

1,615.9 ÷ 4.35 = ____________

279.31 ÷ 4.1 = ____________

363.63 ÷ 3.6 = ____________

123.94 ÷ 3.9 = ____________

34.594 ÷ 4.8 = ____________

561.84 ÷ 79.3 = ____________

355.39 ÷ 5.7 = ____________

Name ______________________________

Decimal Division

Find the quotients.

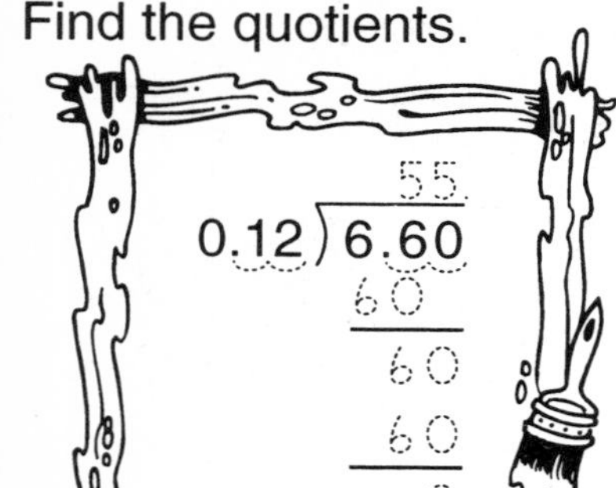

A. $0.4\overline{)26}$ $0.5\overline{)29}$ $0.46\overline{)33.12}$

B. $0.25\overline{)10}$ $3.4\overline{)12.92}$ $0.67\overline{)4.221}$ $0.03\overline{)0.294}$

C. $9.5\overline{)0.8265}$ $0.7\overline{)22.4}$ $0.08\overline{)1.456}$ $2.1\overline{)10.5}$

D. $0.08\overline{)3.68}$ $0.12\overline{)3.12}$ $0.42\overline{)1.470}$ $8.4\overline{)54.6}$

Name ______________________________

Super Shoppers

Solve the problems. Circle your answers.

	Workspace		Workspace
A. Sue bought 3 cans of tomato sauce. How much did she spend?		B. Sean spent $5.00 in all. He bought a package of rolls and some salad. How many pounds of salad did he buy?	
C. Jorge bought a package of rolls and 2 pounds of salad. How much did he spend?		D. Patrick can buy a 5-pound box of spaghetti for $7.50. How much less expensive is this than buying five 1-pound boxes?	
E. Emily spent $8.75 at the salad bar. How much salad did she buy?		F. How many pounds of mushrooms can Judy buy for $9.00?	

Name ______________________________

Powers to the Numbers

Write in exponent form. Then find the value.
You may use a calculator to check your answers.

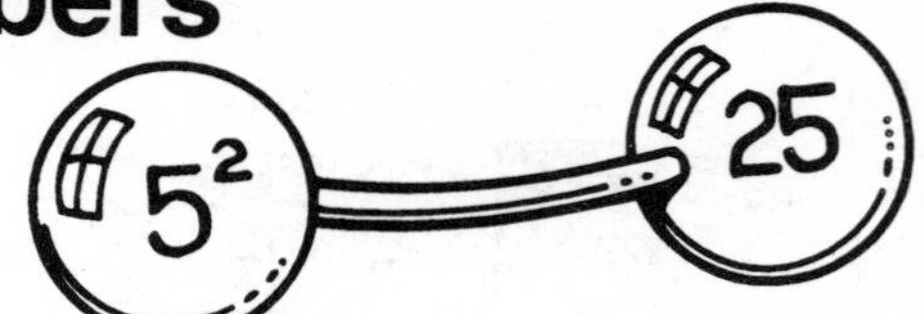

A. five squared

______

______

$8 \cdot 8 \cdot 8 \cdot 8$

______

______

nine to the 4th power

______

______

B. $15 \cdot 15 \cdot 15$

______

______

ten to the 5th power

______

______

seven cubed

______

______

C. one-half cubed

______

______

$4 \cdot 4 \cdot 4 \cdot 4 \cdot 4$

______

______

six to the 3rd power

______

______

D. $2.3 \cdot 2.3 \cdot 2.3$

______

______

twelve squared

______

______

$1.5 \cdot 1.5 \cdot 1.5 \cdot 1.5$

______

______

E. twenty cubed

______

______

$7 \cdot 7 \cdot 7 \cdot 7 \cdot 7 \cdot 7$

______

______

one to the 12th power

______

______

F. $12.3 \cdot 12.3$

______

______

two to the 6th power

______

______

three-fourths squared

______

______

Name ______________________________

Scientific Notation

Scientific notation is an easy way to represent a large number. To write a number in scientific notation, move the decimal point the number of places necessary to get a number from 1 to 9. Then write the number from 1 to 9 with a multiplication sign and a power of 10 showing the number of places you moved the decimal point.

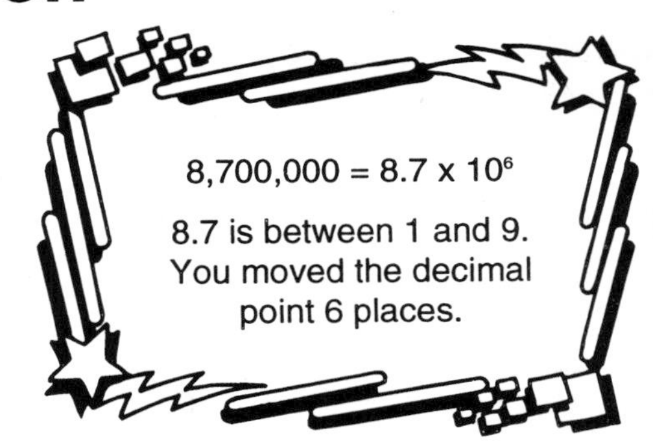

A.	5,300	70,000	260	890,000
	5.3 x 10^3	________	________	________
B.	450	8,200	45,300	976,000
	________	________	________	________
C.	416,000	52,000	93,700	4,960
	________	________	________	________

Write the numbers in standard form.

D.	3.5 x 10^2	4.37 x 10^4	9.7 x 10^5	1.315 x 10^3
	350	________	________	________
E.	3.4 x 10^3	8.37 x 10^5	1.05 x 10^2	9.876 x 10^4
	________	________	________	________
F.	2.46 x 10^6	9.21 x 10^4	3.89 x 10^6	8.4 x 10^8
	________	________	________	________

Name ____________________

Even Steven

A number is divisible by

2	3	4	5	6	9	10
if it ends in 0, 2, 4, 6, or 8	if the sum of the digits is divisible by 3	if the last two digits form a number that is divisible by 4	if it ends in 0 or 5	if it is divisible by 2 and 3	if the sum of the digits is divisible by 9	if it ends in 0

Complete the table. Write **Y** (yes) or **N** (no).

Number	Divisible by						
	2	3	4	5	6	9	10
540							
346							
621							
2,690							
5,211							
4,002							
6,732							
9,017							
10,950							
12,579							
34,782							
56,712							
67,125							
79,470							

Name ______________________________

Primo Primes

A prime number has only two whole-number factors—itself and 1.
A composite number has more than two whole-number factors.
Write **prime** or **composite** beside each number.

A.	3	________	41	________		
B.	18	________	19	________		
C.	64	________	39	________	52	________
D.	10	________	11	________	15	________
E.	6	________	17	________	25	________
F.	5	________	51	________	36	________
G.	49	________	7	________	91	________
H.	53	________	85	________	13	________
I.	89	________	23	________	93	________
J.	43	________	79	________	47	________
K.	73	________	75	________	86	________
L.	67	________	83	________	29	________
M.	31	________	87	________	97	________
N.	94	________	59	________	37	________
O.	61	________	81	________	63	________

Name ______________________________

Perfect Products

List all of the factors of each number from least to greatest. Then tell whether the number of factors is odd or even.

	Number	Factors	Odd or Even Number of Factors
A.	12	1, 2, 3, 4, 6, 12	Even
B.	16		
C.	18		
D.	20		
E.	25		
F.	32		
G.	36		
H.	40		
I.	48		
J.	56		
K.	60		
L.	64		
M.	72		
N.	81		
O.	100		
P.	121		
Q.	144		
R.	225		

Look at the numbers that have an odd number of factors. The middle factor of each number should be the square root of the number.

Name ______________________________

Find the Prime Factors

Draw a factor tree to find the prime factors. Then write the prime factors using exponents.

A.

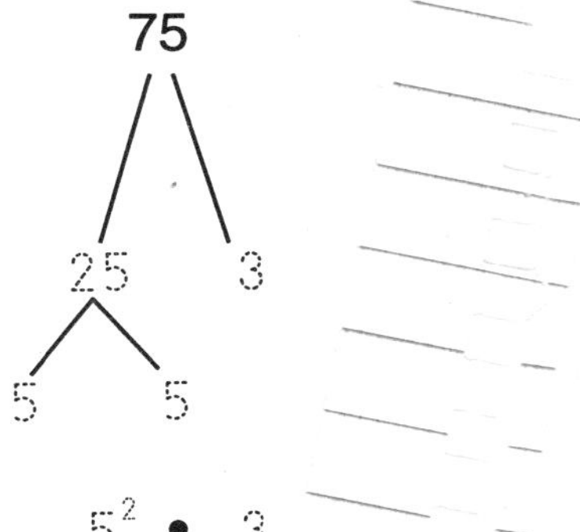

88 ____________

54 ____________

B.

20 ____________

50 ____________

36 ____________

C.

3 ____________

90 ____________

120 ____________

D.

60 ____________

32 ____________

Name ______________________________

Factors Are Great!

On scratch paper, find the factors for each number. Write the **greatest common factor** (GCF) for each pair of numbers on the line below the pair.

Factors of 18 | Factors of 24
- Factors of 18 only: 9, 18
- Common: 1, 2, 3, (6)
- Factors of 24 only: 4, 8, 12, 24

GCF = 6

A.	16 and 40 ______	10 and 21 ______	
B.	24 and 40 ______	36 and 54 ______	48 and 64 ______
C.	18 and 27 ______	12 and 36 ______	21 and 28 ______
D.	16 and 24 ______	18 and 30 ______	8 and 27 ______
E.	45 and 60 ______	28 and 42 ______	48 and 72 ______
F.	26 and 51 ______	100 and 130 ______	24 and 72 ______
G.	27 and 81 ______	18 and 32 ______	42 and 56 ______
H.	90 and 189 ______	91 and 95 ______	84 and 108 ______
I.	144 and 216 ______	136 and 162 ______	121 and 143 ______

Name ____________________

Multiples, at Least

On scratch paper, find the multiples of each number. Write the **least common multiple** (LCM) for each pair of numbers on the line below the pair.

Multiples of 4: 4, 8, 16, 20, 28, 32, ...
Multiples of 6: 6, 18, 30, 42, 48, ...
Both: (12), 24, 36
LCM = 12

A. 5 and 9 ____ 4 and 18 ____

B. 3 and 4 ____ 6 and 21 ____ 18 and 27 ____

C. 4 and 10 ____ 8 and 18 ____ 9 and 36 ____

D. 20 and 25 ____ 18 and 30 ____ 30 and 70 ____

E. 18 and 60 ____ 27 and 36 ____ 20 and 24 ____

Rewrite each pair of fractions using the LCM.

F. $\frac{2}{9}$ and $\frac{4}{15}$ ____ $\frac{3}{4}$ and $\frac{1}{6}$ ____ $\frac{2}{3}$ and $\frac{4}{5}$ ____

G. $\frac{3}{5}$ and $\frac{1}{2}$ ____ $\frac{5}{8}$ and $\frac{3}{4}$ ____ $\frac{2}{3}$ and $\frac{3}{4}$ ____

H. $\frac{3}{7}$ and $\frac{3}{5}$ ____ $\frac{7}{9}$ and $\frac{5}{6}$ ____ $\frac{5}{6}$ and $\frac{3}{8}$ ____

I. $\frac{6}{9}$ and $\frac{7}{8}$ ____ $\frac{2}{5}$ and $\frac{1}{6}$ ____ $\frac{1}{8}$ and $\frac{3}{7}$ ____

Name ______________________________

Order These

Write in order from least to greatest. Use scratch paper if necessary.

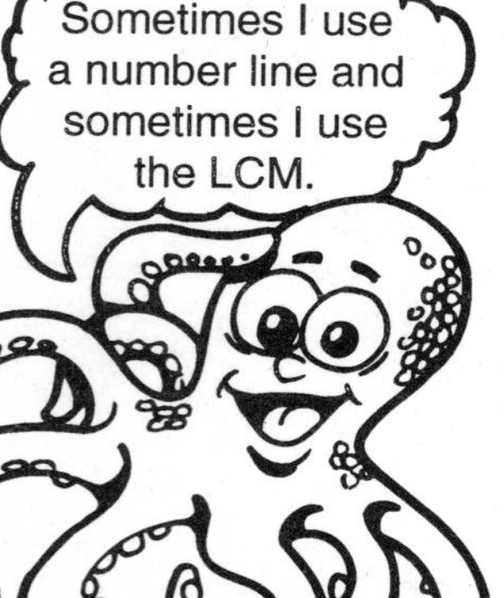

Sometimes I use
a number line and
sometimes I use
the LCM.

A. $\frac{1}{5}, \frac{1}{3}, \frac{1}{4}$ $\frac{4}{9}, \frac{1}{2}, \frac{2}{3}$ $\frac{2}{5}, \frac{7}{15}, \frac{1}{2}$

$\frac{1}{5}, \frac{1}{4}, \frac{1}{3}$ ________   ________   ________

B. $\frac{5}{6}, \frac{6}{7}, \frac{3}{21}$ $\frac{5}{8}, \frac{1}{2}, \frac{3}{4}$ $\frac{4}{5}, \frac{1}{2}, \frac{9}{10}$

________   ________   ________

C. $\frac{2}{5}, \frac{1}{2}, \frac{3}{10}$ $\frac{3}{5}, \frac{1}{2}, \frac{1}{4}$ $\frac{3}{8}, \frac{2}{3}, \frac{1}{4}$ $\frac{7}{15}, \frac{2}{3}, \frac{1}{5}$

________   ________   ________   ________

D. $\frac{3}{4}, \frac{3}{5}, \frac{7}{10}$ $\frac{1}{3}, \frac{3}{8}, \frac{1}{4}$ $\frac{2}{5}, \frac{4}{9}, \frac{11}{15}$ $\frac{5}{6}, \frac{3}{8}, \frac{1}{2}$

________   ________   ________   ________

E. $\frac{2}{5}, \frac{1}{3}, \frac{1}{4}$ $\frac{2}{9}, \frac{5}{18}, \frac{3}{6}$ $\frac{5}{6}, \frac{4}{5}, \frac{3}{4}$ $\frac{1}{9}, \frac{1}{7}, \frac{1}{8}$

________   ________   ________   ________

F. $\frac{2}{3}, \frac{5}{6}, \frac{3}{4}$ $\frac{1}{2}, \frac{3}{10}, \frac{4}{9}$ $\frac{5}{6}, \frac{2}{5}, \frac{2}{3}$ $\frac{2}{7}, \frac{2}{5}, \frac{3}{10}$

________   ________   ________   ________

G. $\frac{5}{6}, \frac{7}{12}, \frac{2}{3}$ $\frac{1}{5}, \frac{1}{9}, \frac{1}{3}$ $\frac{2}{7}, \frac{2}{11}, \frac{2}{5}$ $\frac{1}{3}, \frac{3}{10}, \frac{2}{5}$

________   ________   ________   ________

Name ____________________

Watch the Signs

Add or subtract. Write your answer in the simplest form.

A. $\frac{2}{3} + \frac{1}{3} = \frac{3}{3} = 1$ $\quad \frac{1}{8} + \frac{4}{8}$ $\quad \frac{6}{11} + \frac{2}{11}$ $\quad \frac{10}{13} - \frac{4}{13}$

B. $\frac{11}{12} - \frac{5}{12}$ $\quad \frac{4}{9} + \frac{2}{9}$ $\quad \frac{2}{15} + \frac{8}{15}$ $\quad \frac{3}{20} + \frac{11}{20}$

C. $\frac{2}{5} + \frac{1}{2}$ $\quad \frac{7}{12} + \frac{1}{4}$ $\quad \frac{9}{10} - \frac{1}{2}$ $\quad \frac{2}{3} - \frac{1}{6}$

D. $\frac{13}{18} - \frac{2}{9}$ $\quad \frac{7}{24} + \frac{5}{12}$ $\quad \frac{5}{8} - \frac{4}{7}$ $\quad \frac{14}{15} + \frac{8}{9}$

E. $\frac{4}{9} + \frac{3}{4}$ $\quad \frac{2}{5} + \frac{4}{15}$ $\quad \frac{1}{3} - \frac{1}{7}$ $\quad \frac{4}{5} - \frac{1}{3}$

F. $\frac{1}{9} - \frac{1}{12}$ $\quad \frac{3}{10} + \frac{5}{6} + \frac{2}{5}$ $\quad \frac{4}{9} + \frac{1}{3} + \frac{5}{6}$

Name ______________________________ Adding mixed numbers

Mixed Number Sums

Add. Write your answer in the simplest form.

I have to find the least common denominator first!

A. $3\frac{1}{2} + 2\frac{1}{6}$ $\qquad$ $7\frac{3}{10} + 9\frac{3}{4}$

B. $12\frac{5}{6} + 6\frac{7}{9}$ $\qquad$ $6\frac{4}{7} + 2\frac{9}{14}$ $\qquad$ $8\frac{7}{12} + 6\frac{5}{8}$ $\qquad$ $11\frac{1}{6} + 7\frac{1}{2}$

C. $1\frac{5}{6} + 6\frac{1}{2}$ $\qquad$ $7\frac{2}{3} + 6\frac{3}{5}$ $\qquad$ $4\frac{1}{4} + 7\frac{7}{8}$ $\qquad$ $3\frac{2}{3} + 2\frac{2}{3}$

D. $6\frac{1}{6} + 1\frac{11}{12}$ $\qquad$ $3\frac{3}{4} + 6\frac{1}{2}$ $\qquad$ $2\frac{5}{6} + 5\frac{3}{4}$ $\qquad$ $2\frac{1}{8} + 8\frac{2}{3}$

E. $3\frac{1}{3}$, $4\frac{5}{6}$, $+ 1\frac{1}{12}$ $\qquad$ $6\frac{2}{3}$, $1\frac{1}{3}$, $+ 3\frac{1}{2}$ $\qquad$ $5\frac{1}{2}$, $2\frac{1}{3}$, $+ 11\frac{1}{6}$ $\qquad$ $2\frac{1}{2}$, $1\frac{5}{8}$, $+ 3\frac{3}{4}$

F. $5\frac{1}{4}$, $9\frac{1}{12}$, $+ 4\frac{1}{6}$ $\qquad$ $2\frac{1}{15}$, $11\frac{2}{5}$, $+ 3\frac{1}{3}$ $\qquad$ $7\frac{1}{2}$, $14\frac{1}{10}$, $+ 4\frac{2}{5}$ $\qquad$ $3\frac{2}{3}$, $1\frac{1}{2}$, $+ 4\frac{3}{4}$

Name ____

Mixed Number Differences

Subtract. Write your answer in the simplest form.

A.	$8\frac{4}{5} - 4\frac{1}{2}$	$7\frac{3}{10} - 6\frac{3}{4}$	$9\frac{7}{9} - 5\frac{1}{3}$	$8\frac{3}{4} - 2\frac{5}{8}$
B.	$6\frac{5}{6} - 3\frac{7}{9}$	$2\frac{1}{9} - \frac{1}{2}$	$18\frac{1}{3} - 9$	$17\frac{3}{5} - 6\frac{1}{3}$
C.	$4\frac{9}{10} - 3\frac{2}{5}$	$4\frac{5}{8} - 1\frac{1}{4}$	$4\frac{7}{9} - 2\frac{4}{9}$	$3\frac{3}{10} - 2$
D.	$5\frac{9}{10} - 4\frac{3}{5}$	$3 - 1\frac{7}{10}$	$6\frac{3}{8} - 2\frac{3}{16}$	$4\frac{1}{10} - 2\frac{3}{10}$
E.	$8\frac{1}{4} - 4\frac{3}{4}$	$12\frac{2}{5} - 8\frac{9}{10}$	$15\frac{1}{2} - 9\frac{9}{16}$	$12\frac{9}{10} - \frac{17}{20}$
F.	$4\frac{1}{2} - 2\frac{3}{4}$	$14\frac{1}{9} - 9\frac{2}{3}$	$9\frac{1}{3} - 6\frac{1}{2}$	

Did you remember to find the least common denominator? Did you regroup when you needed to?

Name ______________________________

Fraction Practice

Add or subtract. Write your answers in the simplest form.

A.

$\frac{1}{5} + \frac{3}{10}$

$\frac{3}{4} - \frac{3}{8}$

$4\frac{7}{8} + 4\frac{7}{8}$

B.

$\frac{4}{5} + 1\frac{5}{6}$

$9\frac{1}{8} + 6\frac{3}{4}$

$16\frac{5}{11} - 9$

$6\frac{4}{7} + 2\frac{1}{5}$

C.

$13\frac{1}{3} + 9\frac{8}{17}$

$7 - 6\frac{1}{8}$

$8\frac{1}{12} + 3\frac{1}{3}$

$5\frac{9}{14} - 3\frac{6}{7}$

D.

$6 - 1\frac{19}{21}$

$6\frac{7}{11} + 3$

$4\frac{1}{2} + 10\frac{11}{14}$

$9\frac{3}{7} - 4\frac{1}{6}$

E.

$\frac{5}{12} + 4\frac{5}{6}$

$15\frac{1}{7} - 6\frac{2}{3}$

$5\frac{9}{10} + 1\frac{1}{2}$

$9\frac{1}{4} - 3\frac{2}{5}$

F.

$\frac{11}{14} + 1\frac{1}{7} + 3\frac{1}{4}$

$6\frac{1}{6} + 5\frac{1}{3} + 10\frac{1}{2}$

$19\frac{1}{2} - 9\frac{3}{4}$

$2\frac{3}{8} + 5\frac{1}{6} + 3\frac{11}{12}$

Name ____________________

Products From Fractions

Multiply. Divide any numerator and denominator by a common factor to make the fractions easier to multiply. Write your answers in the simplest form.

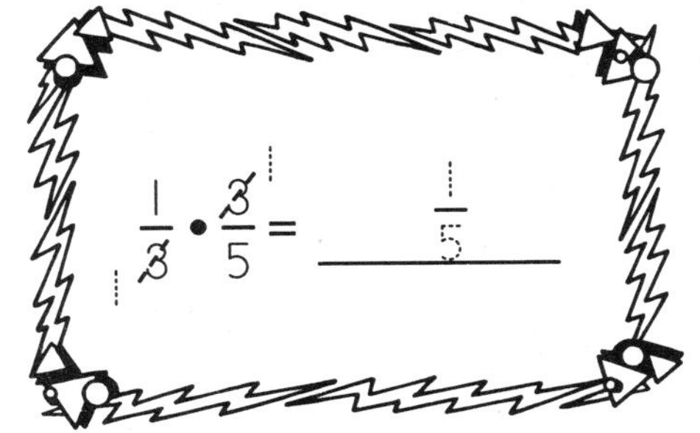

A. $\frac{1}{4} \bullet \frac{4}{9} =$ ______ $\frac{7}{8} \bullet \frac{20}{21} =$ ______

B. $\frac{7}{8} \bullet \frac{8}{7} =$ ______ $\frac{9}{16} \bullet \frac{10}{9} =$ ______

C. $\frac{7}{11} \bullet \frac{23}{42} =$ ______ $\frac{3}{8} \bullet \frac{4}{5} =$ ______ $\frac{5}{3} \bullet \frac{2}{5} =$ ______

D. $\frac{3}{2} \bullet \frac{5}{6} =$ ______ $\frac{3}{4} \bullet \frac{2}{9} =$ ______ $\frac{2}{3} \bullet \frac{6}{5} =$ ______

E. $\frac{5}{7} \bullet \frac{7}{10} =$ ______ $\frac{4}{5} \bullet \frac{1}{8} =$ ______ $\frac{7}{12} \bullet \frac{3}{7} =$ ______

F. $\frac{4}{7} \bullet \frac{7}{2} =$ ______ $\frac{3}{5} \bullet \frac{25}{6} =$ ______ $\frac{4}{5} \bullet \frac{3}{2} =$ ______

G. $\frac{7}{8} \bullet \frac{4}{21} =$ ______ $\frac{2}{11} \bullet \frac{11}{24} =$ ______ $\frac{1}{3} \bullet \frac{3}{8} =$ ______

H. $\frac{3}{5} \bullet \frac{2}{3} =$ ______ $\frac{6}{7} \bullet \frac{5}{12} =$ ______ $\frac{4}{5} \bullet \frac{2}{7} =$ ______

I. $\frac{7}{10} \bullet \frac{5}{8} =$ ______ $\frac{2}{7} \bullet \frac{7}{8} =$ ______ $\frac{6}{7} \bullet \frac{14}{15} =$ ______

J. $\frac{3}{4} \bullet \frac{8}{9} =$ ______ $\frac{8}{5} \bullet \frac{15}{16} =$ ______ $\frac{9}{10} \bullet \frac{4}{3} =$ ______

K. $\frac{5}{2} \bullet \frac{1}{2} =$ ______ $\frac{5}{3} \bullet \frac{15}{8} =$ ______ $\frac{4}{3} \bullet \frac{5}{8} =$ ______

L. $\frac{1}{3} \bullet \frac{3}{5} \bullet \frac{5}{7} =$ ______ $\frac{5}{2} \bullet \frac{2}{3} \bullet \frac{3}{10} =$ ______ $\frac{3}{4} \bullet \frac{4}{5} \bullet \frac{1}{3} =$ ______

Mixed Number Multiplication

Find the products. Rewrite mixed numbers as improper fractions before you multiply. Write your answers in the simplest form.

A. 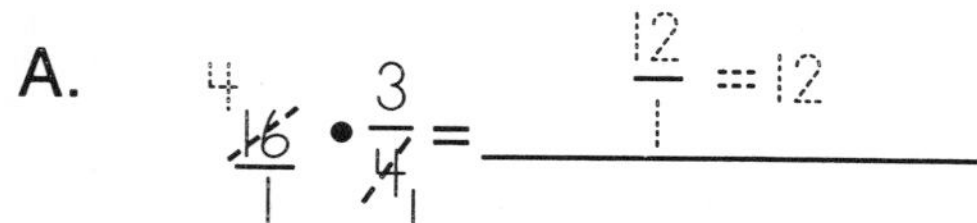 $16 \bullet \frac{3}{4} =$ ________ $20 \bullet \frac{3}{5} =$ ________

B. $6\frac{2}{3} \bullet \frac{1}{4} =$ ________ $5\frac{1}{8} \bullet 6 =$ ________ $3 \bullet 5\frac{1}{2} =$ ________

C. $7\frac{1}{7} \bullet \frac{3}{8} =$ ________ $7\frac{3}{4} \bullet 20 =$ ________ $4\frac{3}{8} \bullet \frac{2}{5} =$ ________

D. $1\frac{1}{3} \bullet 30 =$ ________ $2\frac{1}{2} \bullet \frac{5}{6} =$ ________ $9\frac{2}{9} \bullet \frac{18}{25} =$ ________

E. $\frac{3}{8} \bullet 7\frac{2}{3} =$ ________ $5\frac{1}{4} \bullet 16 =$ ________ $7\frac{1}{3} \bullet \frac{6}{11} =$ ________

F. $2\frac{2}{3} \bullet 4\frac{1}{2} =$ ________ $9\frac{7}{8} \bullet 2\frac{2}{3} =$ ________ $12\frac{1}{2} \bullet \frac{4}{5} =$ ________

G. $4\frac{2}{11} \bullet 22 =$ ________ $18 \bullet 7\frac{4}{9} =$ ________ $\frac{1}{2} \bullet 24 =$ ________

H. $4\frac{2}{5} \bullet 25 =$ ________ $5\frac{1}{3} \bullet 9\frac{1}{8} =$ ________ $15 \bullet 8\frac{1}{3} =$ ________

I. $15 \bullet 9\frac{2}{3} =$ ________ $2\frac{1}{4} \bullet 11\frac{1}{3} =$ ________ $5\frac{1}{3} \bullet 1\frac{1}{4} =$ ________

J. $2\frac{3}{4} \bullet \frac{1}{8} =$ ________ $3\frac{1}{3} \bullet 4\frac{2}{5} =$ ________ $2\frac{5}{8} \bullet 16 =$ ________

K. $3\frac{3}{4} \bullet 16 =$ ________ $\frac{1}{9} \bullet 3\frac{2}{3} =$ ________ $6\frac{2}{3} \bullet 24 =$ ________

Name ______________________________

How Many Equal Parts?

To divide fractions, rewrite the problem and multiply by the reciprocal of the divisor. Circle each quotient in the simplest form.

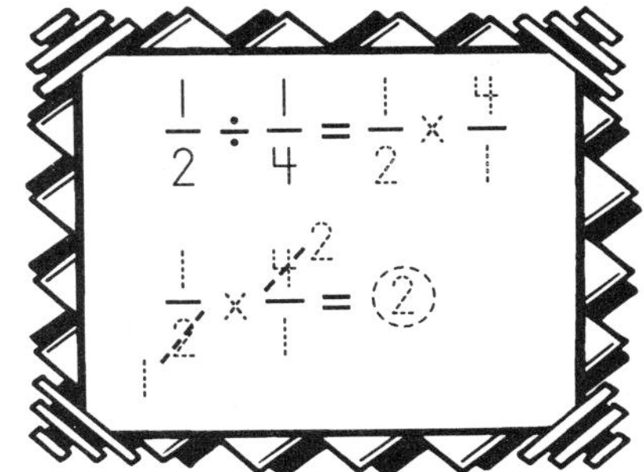

A. $\frac{8}{9} \div \frac{4}{5} =$ ________ $\frac{7}{8} \div \frac{7}{8} =$ ________

B. $\frac{1}{2} \div \frac{3}{4} =$ ________ $\frac{5}{6} \div \frac{5}{12} =$ ________

C. $\frac{3}{4} \div \frac{1}{2} =$ ________ $\frac{2}{3} \div \frac{1}{3} =$ ________ $\frac{7}{12} \div \frac{1}{12} =$ ________

D. $\frac{4}{5} \div \frac{1}{8} =$ ________ $\frac{1}{3} \div \frac{1}{12} =$ ________ $\frac{5}{2} \div \frac{3}{10} =$ ________

E. $\frac{4}{5} \div \frac{7}{5} =$ ________ $\frac{3}{4} \div \frac{1}{8} =$ ________ $\frac{3}{8} \div \frac{1}{10} =$ ________

F. $\frac{1}{3} \div \frac{1}{2} =$ ________ $\frac{1}{3} \div \frac{1}{9} =$ ________ $\frac{1}{9} \div \frac{1}{3} =$ ________

G. $\frac{5}{8} \div \frac{3}{4} =$ ________ $\frac{7}{10} \div \frac{2}{3} =$ ________ $\frac{1}{10} \div \frac{1}{5} =$ ________

H. $\frac{5}{6} \div \frac{2}{3} =$ ________ $\frac{3}{10} \div \frac{1}{5} =$ ________ $\frac{1}{2} \div \frac{3}{10} =$ ________

I. $\frac{5}{8} \div \frac{1}{4} =$ ________ $\frac{7}{2} \div \frac{3}{4} =$ ________ $\frac{5}{4} \div \frac{1}{8} =$ ________

J. $\frac{5}{2} \div \frac{3}{8} =$ ________ $\frac{3}{10} \div \frac{9}{10} =$ ________ $\frac{1}{5} \div \frac{7}{10} =$ ________

K. $\frac{3}{10} \div \frac{2}{5} =$ ________ $\frac{3}{5} \div \frac{1}{4} =$ ________ $\frac{3}{8} \div \frac{3}{4} =$ ________

Mixed Number Division

Change the mixed numbers to improper fractions. Then multiply the first fraction by the reciprocal of the second fraction. Circle each answer in the simplest form.

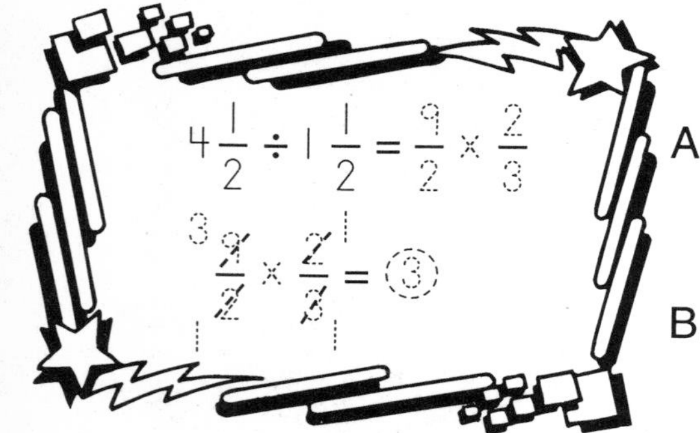

A. $4\frac{2}{3} \div 2 =$ ________ $4\frac{3}{4} \div 2\frac{3}{4} =$ ________

B. $6\frac{1}{4} \div 12\frac{1}{2} =$ ________ $12\frac{4}{5} \div 8 =$ ________

C. $7\frac{5}{6} \div 1\frac{5}{6} =$ ________ $5\frac{1}{4} \div 1\frac{1}{6} =$ ________ $6 \div 1\frac{2}{3} =$ ________

D. $15 \div 1\frac{1}{2} =$ ________ $6\frac{1}{4} \div 5 =$ ________ $16\frac{1}{3} \div 4\frac{2}{3} =$ ________

E. $\frac{2}{3} \div 1\frac{1}{3} =$ ________ $18 \div 7\frac{1}{5} =$ ________ $8\frac{1}{4} \div 1\frac{3}{8} =$ ________

F. $1\frac{1}{2} \div \frac{3}{8} =$ ________ $2\frac{4}{7} \div 2 =$ ________ $3\frac{1}{3} \div \frac{5}{6} =$ ________

G. $1\frac{1}{2} \div 4\frac{1}{2} =$ ________ $3\frac{1}{4} \div 1\frac{3}{8} =$ ________ $\frac{7}{8} \div 3\frac{1}{2} =$ ________

H. $2\frac{1}{2} \div 1\frac{1}{2} =$ ________ $10 \div 3\frac{1}{3} =$ ________ $4\frac{2}{5} \div 4 =$ ________

I. $2\frac{4}{5} \div 6\frac{2}{3} =$ ________ $2\frac{2}{3} \div 1\frac{1}{6} =$ ________ $5\frac{1}{6} \div 5\frac{2}{3} =$ ________

J. $2\frac{3}{4} \div 5\frac{1}{8} =$ ________ $9\frac{3}{5} \div 2 =$ ________ $6\frac{1}{3} \div 3\frac{1}{5} =$ ________

K. $9 \div 3\frac{1}{3} =$ ________ $1\frac{1}{10} \div \frac{3}{5} =$ ________ $6\frac{3}{4} \div 1\frac{1}{5} =$ ________

Georgina's Famous Chili

Here are the ingredients Georgina uses in her chili.

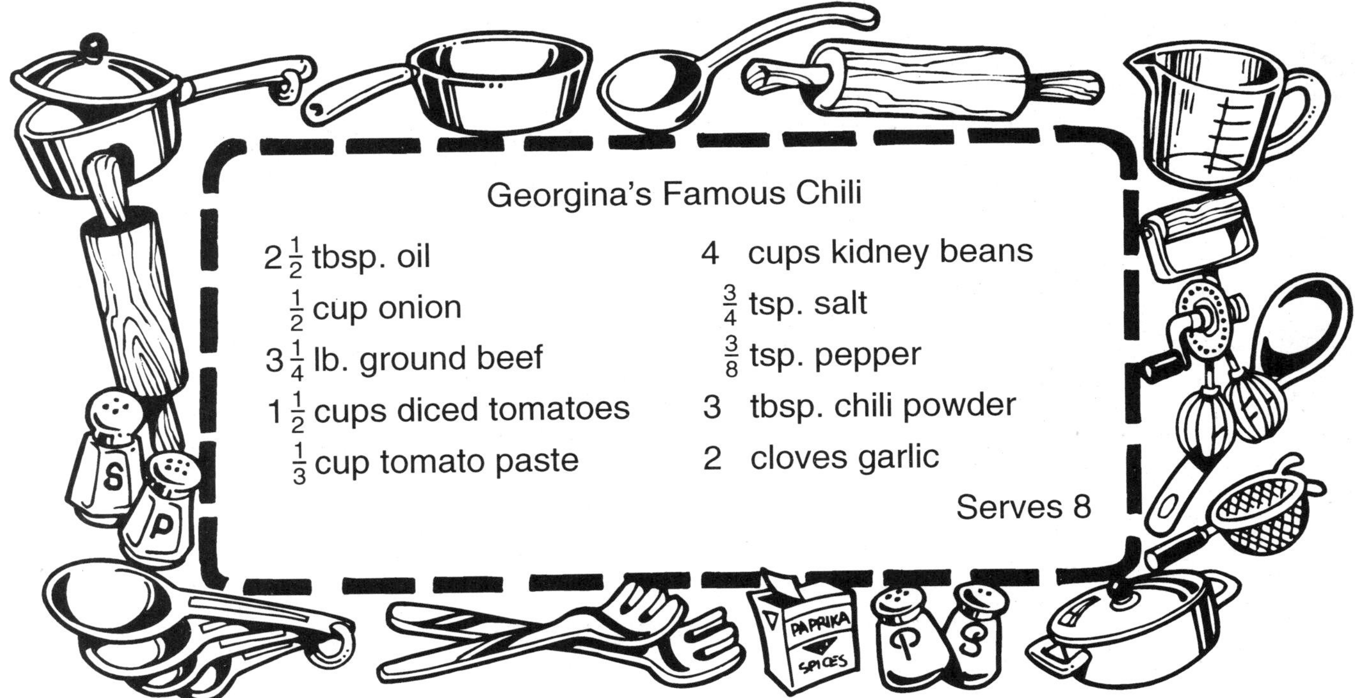

Georgina's Famous Chili

$2\frac{1}{2}$ tbsp. oil	4 cups kidney beans
$\frac{1}{2}$ cup onion	$\frac{3}{4}$ tsp. salt
$3\frac{1}{4}$ lb. ground beef	$\frac{3}{8}$ tsp. pepper
$1\frac{1}{2}$ cups diced tomatoes	3 tbsp. chili powder
$\frac{1}{3}$ cup tomato paste	2 cloves garlic

Serves 8

Revise the amount of each ingredient to serve the number of people shown.

Ingredient	16 servings	4 servings	20 servings	6 servings
oil	5 tbsp.	$1\frac{1}{4}$ tbsp.	$6\frac{1}{4}$ tbsp.	$1\frac{7}{8}$ tbsp.
onion				
ground beef				
tomatoes				
tomato paste				
kidney beans				
salt				
pepper				
chili powder				
garlic				

Name ____

How Many Vowels?

Complete the frequency distribution table by tallying the vowels in the following paragraph. (The cumulative frequency is the sum of the frequency and all the frequencies above it on the table.)

> The alphabet consists of 26 letters. Five of them are vowels. Some vowels are used more often than others. Which do you think is used most in this paragraph?

Vowels Used in Everyday Writing

Vowel	Tally	Frequency	Cumulative Frequency

Use the frequency table to answer the questions.

A. How many vowels were used in the paragraph?

____

B. How many more times was **o** used than **u**?

____

C. Altogether, how many times were **a** and **e** used in the paragraph?

____

D. Was **e** used more often or less often than the other vowels combined?

____

E. Write a question that can be answered by reading the frequency table. Then answer it.

____

____

Name ____

Picture This

Make a pictograph for the data shown in the table below. Write a title on the line above the graph and make a symbol key. List the sports along the left side of the graph and use symbols to indicate the number of people who participate in each.

Mountain Climbing	60 people
Walking	120 people
Running	75 people
Swimming	30 people
Bicycling	45 people
Aerobics	15 people

____

key

Use your pictograph to answer the questions.

A. Why did you choose your title?

____

B. What does each symbol represent? Why did you choose that number value?

____

C. Write something you know from reading your picture graph.

____

____

Name ______________________________

Bars for Cars

Make a double bar graph for the data shown at the right. List the autos along the left side of the graph. For each car, show the number sold in March with one color and the number sold in April with another color. Write a title on the line above the graph and fill in the color key.

	March	April
Compact	2,000	2,300
2-Door Sports	1,200	1,800
4-Door Sedan	1,100	950
Station Wagon	900	850
Van	1,750	2,300

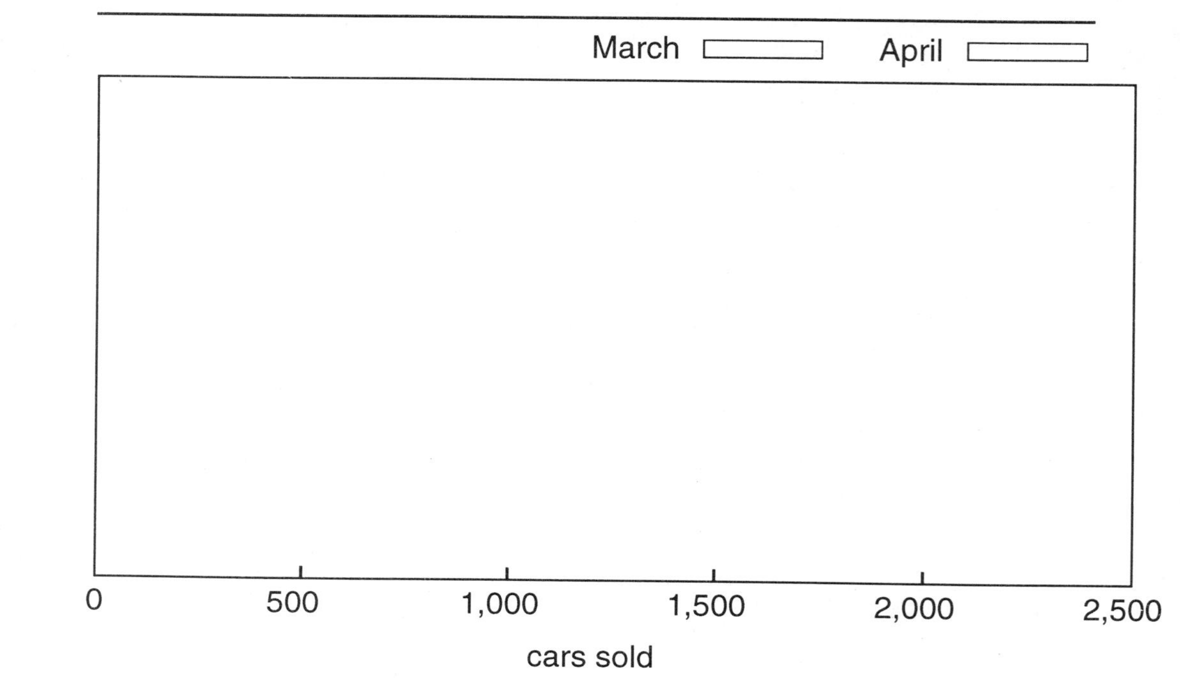

Use the information given to answer the questions.

A. Which car sales did not increase between March and April?

____

B. Which car style had the largest increase in sales?

____

C. How many more compact cars were sold in April than in March?

____

D. Which two car styles had the same sales in April?

____

E. Write something you know from reading the double bar graph.

____

____

Name ______________________________

Video Rental Records

Make a line graph for the data shown on the chart at the right. Write a title on the line above the graph. Put labels along the horizontal and vertical axes. Make a dot to show the number of videos rented each day. Connect the dots with lines.

Day of Week	Number of Videos
Monday	60
Tuesday	48
Wednesday	60
Thursday	72
Friday	108
Saturday	120
Sunday	96

____

Use your line graph to answer the questions.

A. On which day were video rentals highest? Lowest?

____

B. On which two days were video rentals the same?

____

C. Did video rentals increase or decrease between Tuesday and Wednesday?

____

D. What does the graph tell about the trend of video rentals during the course of the week?

____

____

Heavyweight Histogram

Study the histogram below. Then answer the questions.

Weights of Junior League Football Players

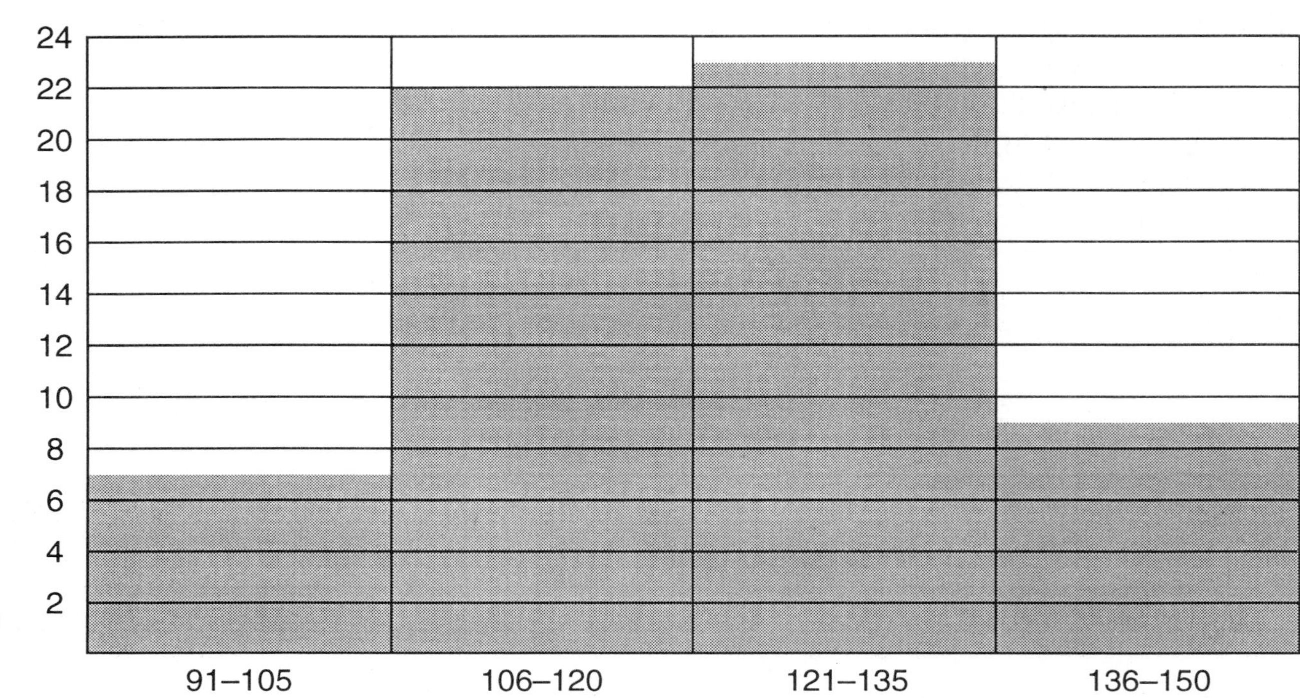

Weight in Pounds

A. What does the width of each bar on the histogram represent? ______________________________

____

B. What does the height of each bar on the histogram represent? ______________________________

____

C. How many players weigh more than 135 pounds? ______________________________

D. What is the weight range with the greatest frequency? ______________________________

E. How many players weigh less than 106 pounds? ______________________________

F. Write something you know from reading the histogram. ______________________________

____

Name ______________________________

Measures of Central Tendency

Eleven students from each math class competed in a math competition. Their scores are shown below.

Teacher	Scores
Ms. Rowe	79, 83, 96, 75, 100, 80, 91, 87, 72, 86, 79
Mrs. Midgely	86, 89, 93, 86, 95, 82, 77, 86, 95, 98, 86
Mr. Maynard	68, 95, 72, 100, 82, 85, 72, 73, 68, 72, 80
Mr. Arnaiz	80, 75, 78, 80, 92, 66, 70, 78, 68, 90, 78
Ms. Silver	73, 68, 75, 82, 69, 85, 75, 78, 75, 88, 78
Ms. Choi	94, 90, 85, 87, 72, 79, 86, 95, 94, 98, 89

range—the difference between the least and greatest numbers

mean—the average

median—the middle number in a set of data

mode—the number that occurs most frequently in a set of data

Find the range, the mean to the nearest tenth, the median, and the mode for each class. Write them on the chart below.

Teacher	Range	Mean	Median	Mode
Ms. Rowe				
Mrs. Midgely				
Mr. Maynard				
Mr. Arnaiz				
Ms. Silver				
Ms. Choi				

Use your data to answer the questions.

A. Whose class had the highest mean?

____

B. Whose class had the smallest range?

____

C. Whose class had a five-point difference between the median and the mode?

____

D. Whose class had the lowest median?

____

Name ____

Box-and-Whisker Graphs

A box-and-whisker graph organizes data and helps you interpret it. Study the box-and-whisker graph shown below. The **median** is the middle number in the ordered data. The **first quartile** is the median of the lower half of the data. The **third quartile** is the median of the upper half of the data.

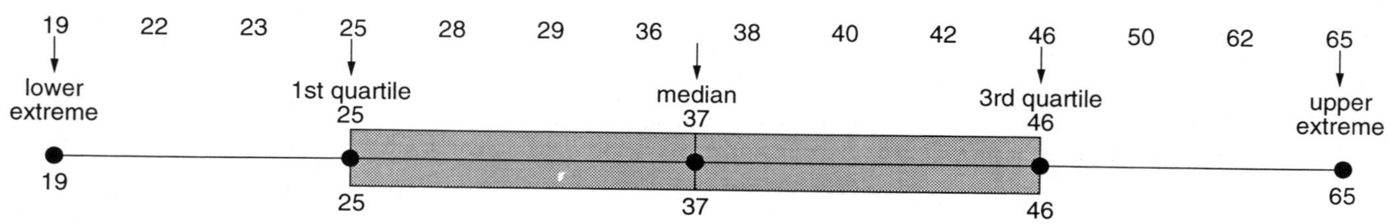

Answer the following questions about the box-and-whisker graph shown at the right.

A. What is the lower extreme? ____________

B. What is the first quartile? ____________

C. What is the median? ____________

D. What is the third quartile? ____________

E. What is the upper extreme? ____________

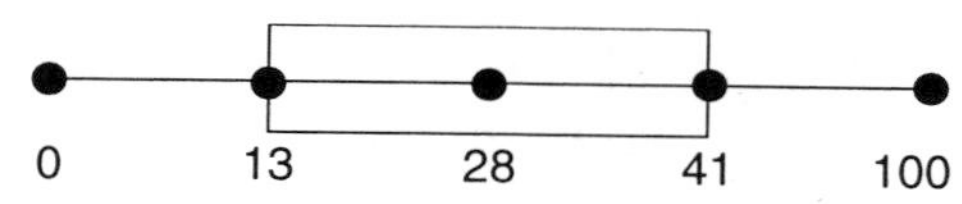

Study the unfinished box-and-whisker graph below. Then answer the questions and record the information on the box-and-whisker graph.

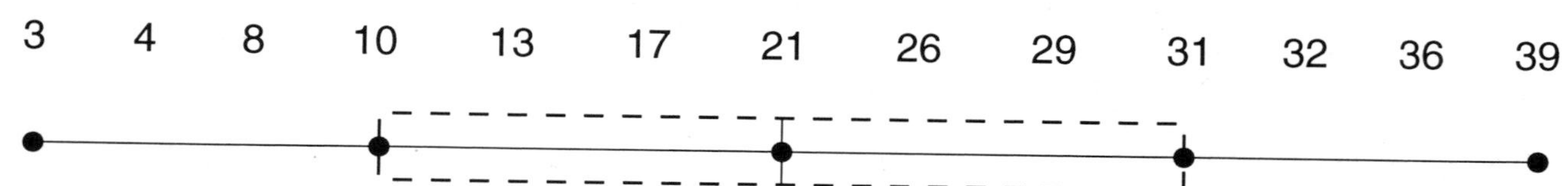

F. What is the lower extreme? ____________

G. What is the first quartile? ____________

H. What is the median? ____________

I. What is the third quartile? ____________

J. What is the upper extreme? ____________

Name ____

Simplest Ratios

Write each ratio as a fraction in its simplest form.

A. 15 to 5 = $\frac{15}{5} = \frac{3}{1}$ $\frac{12}{10}$ = ________

B. 24 : 16 = ________ $\frac{8}{14}$ = ________ 11 to 88 = ________

C. $\frac{26}{10}$ = ________ 9 to 27 = ________ 3 : 33 = ________

D. 10 : 34 = ________ 16 to 36 = ________ $\frac{30}{18}$ = ________

E. $\frac{36}{24}$ = ________ 7 to 35 = ________ $\frac{16}{10}$ = ________

F. 8 : 64 = ________ 28 : 7 = ________ 10 to 90 = ________

G. 64 to 36 = ________ 48 : 144 = ________ $\frac{10}{36}$ = ________

H. 24 : 60 = ________ $\frac{18}{36}$ = ________ 60 to 200 = ________

I. 15 to 18 = ________ $\frac{20}{45}$ = ________ 14 : 28 = ________

J. $\frac{10}{14}$ = ________ 25 to 45 = ________ 24 : 80 = ________

K. 40 to 60 = ________ 15 to 100 = ________ 25 : 75 = ________

L. 5 to 10 = ________ $\frac{6}{8}$ = ________ 84 : 48 = ________

Name ____________________

Rate Per Unit

A **rate** is a ratio that compares quantities of different units. A **unit rate** is a ratio that has 1 as the second term. Write the unit rate for each rate listed below.

A. 48 people in 6 vans ____8 in 1____ | 150 kilometers in 3 hours ________ | 45 cookies in 9 packages ________

B. 30 pounds in 6 weeks ________ | 1,250 words in 5 minutes ________ | 28 days in 4 weeks ________

C. 90 people in 15 cars ________ | 100 cards in 4 packages ________ | 78 centimeters in 3 seconds ________

D. 48 months in 4 years ________ | 144 markers in 6 boxes ________ | 143 players on 11 teams ________

E. 550 miles in 10 hours ________ | 720 minutes in 4 trips ________ | 498 milliliters in 2 glasses ________

F. 258 kilometers in 3 hours ________ | 300 people in 6 buses ________ | 600 flowers in 150 corsages ________

G. 160 pages in 4 hours ________ | 266 rides in 7 weeks ________ | 78 miles in 2 hours ________

H. 210 sit-ups in 6 days ________ | 294 minutes for 7 lessons ________ | 40 apples for 5 children ________

I. 960 miles in 8 hours ________ | 152 crayons in 8 packages ________ | 560 calories in 5 apples ________

Name ______________________________

Unit Pricing

Find the unit prices. Round to the nearest cent if necessary.

A.	3 for \$0.96	7 for \$1.61	
	________	________	
B.	9 for \$0.54	4 for \$1.76	5 for \$1.55
	________	________	________
C.	5 for \$19.95	6 for \$1.41	$1\frac{1}{2}$ for \$3.45
	________	________	________
D.	12 for \$13.32	4 for \$5.00	3 for \$2.91
	________	________	________
E.	3 for \$228.00	20 for \$45.00	8 for \$4.68
	________	________	________
F.	$2\frac{1}{4}$ for \$1.00	8 for \$10.00	5 for \$67.00
	________	________	________
G.	5 for \$18.75	11 for \$1.21	16 for \$8.80
	________	________	________
H.	3 for \$4.74	4 for \$3.00	18 for \$50.00
	________	________	________
I.	25 for \$20.00	5 for \$49.75	6 for \$0.88
	________	________	________
J.	3 for \$0.69	$4\frac{1}{2}$ for \$90.00	23 for \$57.50
	________	________	________

Name ____ Comparing proportions

Bargain Shopping

Find the unit price for the items in each pair. Round to the nearest cent if necessary. Circle the best value in each pair.

A.

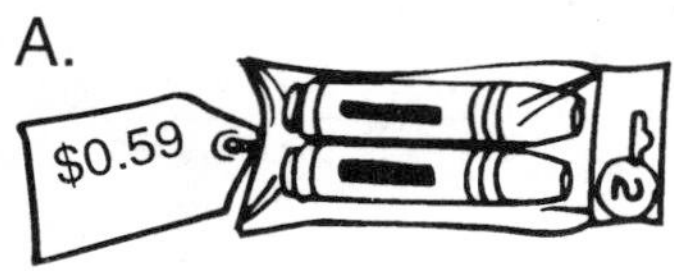

__________ __________

B.

COOKIES
12 oz
$2.55

COOKIES
18 oz.
$3.29

__________ __________

C.

__________ __________

D.

__________ __________

E.

__________ __________

F.

__________ __________

G.

__________ __________

H.

__________ __________

I.

__________ __________

J.

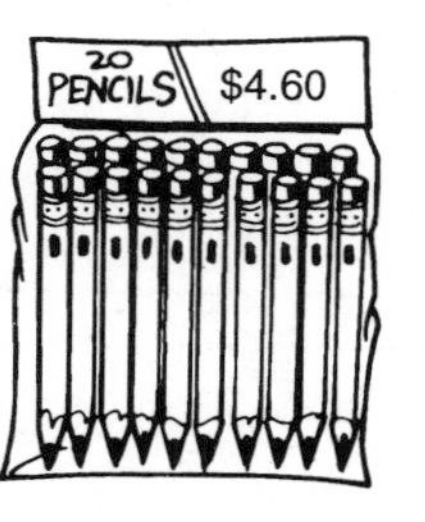

__________ __________

Name ______________________________

Same Shape but Different Dimensions

Write the measurements of each pair of figures as a proportion to find the missing number.

A.

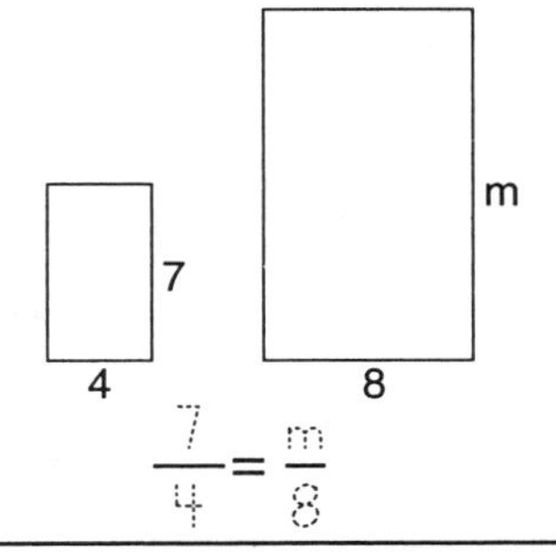

$\frac{7}{4} = \frac{m}{8}$

$56 = 4m$

$m = 14$

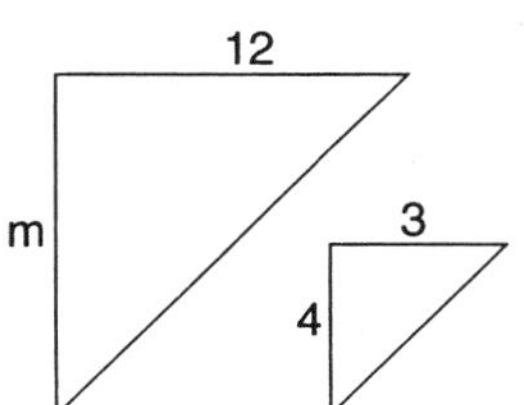

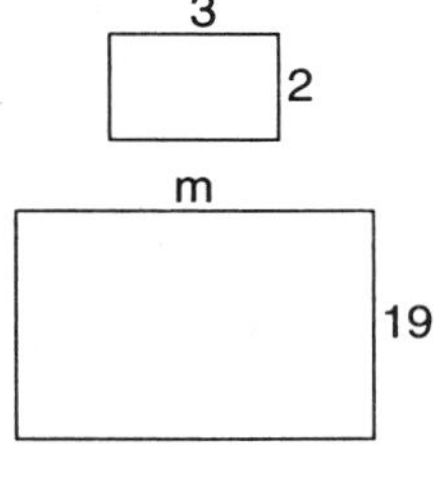

B.

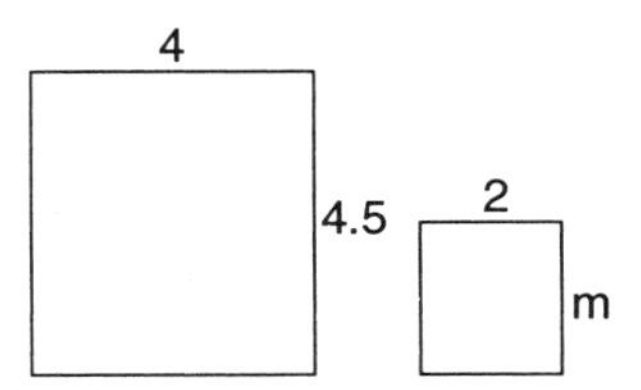

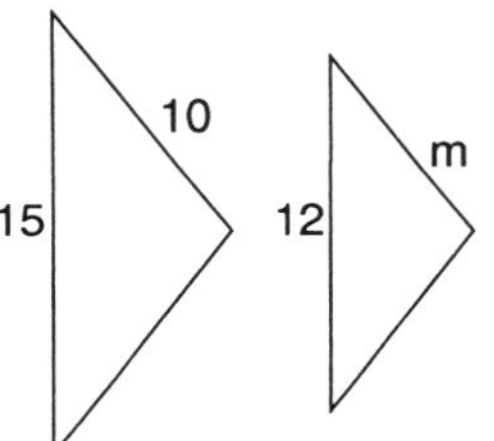

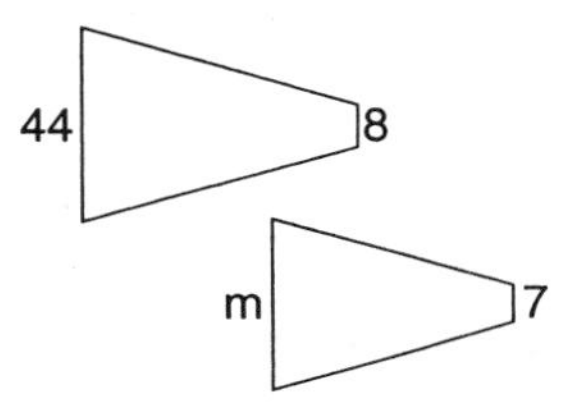

C.

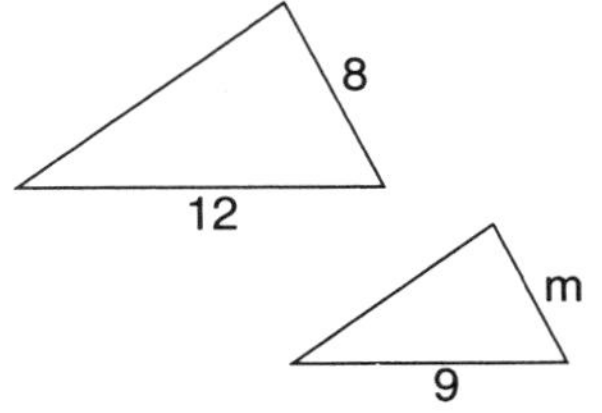

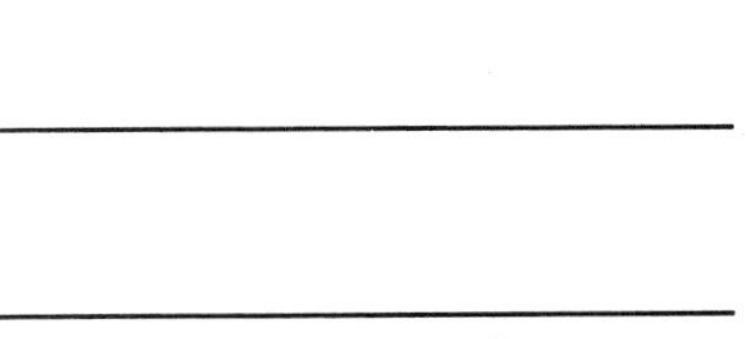

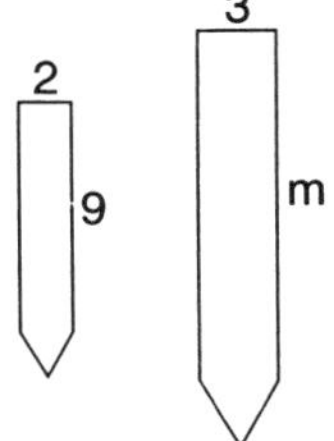

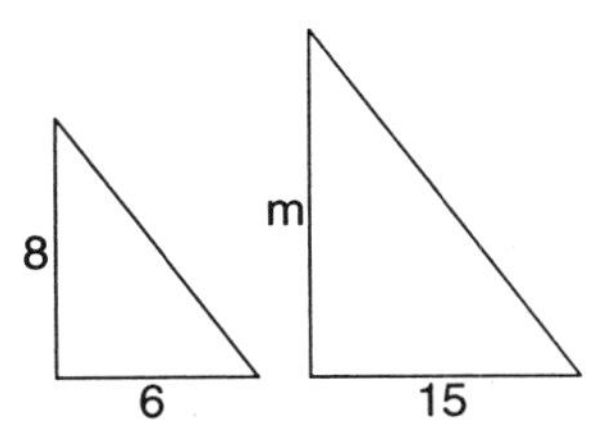

Writing Percents

Write each decimal or fraction as a percent.

Percent
means
"out of 100."

A.	$\frac{35}{100}$ = ______	$\frac{17}{100}$ = ______	0.93 = ______	
B.	0.41 = ______	0.45 = ______	0.91 = ______	
C.	0.23 = ______	$\frac{75}{100}$ = ______	0.72 = ______	0.51 = ______
D.	$\frac{81}{100}$ = ______	0.01 = ______	$\frac{25}{100}$ = ______	$\frac{73}{100}$ = ______
E.	$\frac{27}{100}$ = ______	$\frac{1}{2}$ = ______	$\frac{43}{100}$ = ______	0.03 = ______
F.	0.31 = ______	0.1 = ______	$\frac{4}{10}$ = ______	0.85 = ______
G.	0.05 = ______	$\frac{39}{100}$ = ______	$\frac{6}{10}$ = ______	$\frac{29}{100}$ = ______
H.	$\frac{9}{100}$ = ______	$\frac{8}{10}$ = ______	0.08 = ______	0.15 = ______
I.	0.63 = ______	0.2 = ______	$\frac{45}{100}$ = ______	0.4 = ______
J.	0.71 = ______	0.86 = ______	$\frac{23}{100}$ = ______	0.07 = ______
K.	$\frac{97}{100}$ = ______	$\frac{7}{100}$ = ______	$\frac{7}{10}$ = ______	0.98 = ______
L.	0.42 = ______	$\frac{9}{10}$ = ______	$\frac{33}{100}$ = ______	0.66 = ______
M.	0.31 = ______	$\frac{99}{100}$ = ______	$\frac{47}{100}$ = ______	0.02 = ______

Name ____________________

Missing Percents

Set up a proportion to find each missing percent. Circle the percent.

What percent of 20 is 3?

$\frac{n}{100} = \frac{3}{20}$

$20n = 300$

$n = 15 \quad \frac{15}{100} = 15\%$

A. What percent of 55 is 22?

B. What percent of 25 is 18?

What percent of 50 is 34?

C. What percent of 75 is 45?

What percent of 78 is 39?

D. 15 is what percent of 75?

18 is what percent of 60?

E. What percent of 40 is 8?

3 is what percent of 24?

F. 17 is what percent of 68?

What percent of 360 is 90?

Name ____________________

Missing Numbers

Set up a proportion to find each missing number. Circle the number.

2 is 10% of what number?

$$\frac{2}{n} = \frac{10}{100}$$

$$10n = 200$$

$$n = 20$$

A. 15% of what number is 75?

B. 20 is 4% of what number?

75% of what number is 24?

C. 60 is 20% of what number?

40 is 25% of what number?

D. 15 is 6% of what number?

25% of what number is 15?

E. 6 is 50% of what number?

270 is 54% of what number?

F. 150% of what number is 105?

9% of what number is 54?

Name ______________________________

Solve It in Order

Find the value of each expression. Follow the order of operations shown on the banner.

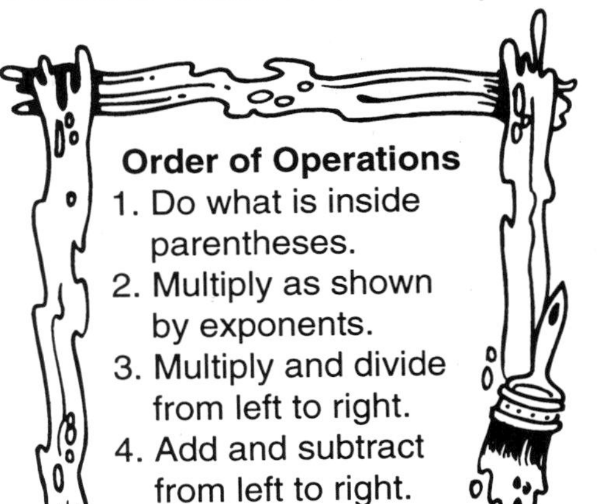
Order of Operations
1. Do what is inside parentheses.
2. Multiply as shown by exponents.
3. Multiply and divide from left to right.
4. Add and subtract from left to right.

A.		$16 - 7 \bullet 2$	$12 + 9 \bullet 3 - 28$
B.		$54 + 24 \div 3 - 30$	$9 \bullet 4 \div 2$
C.		$8 \bullet 2 + 45 \div 9$	$16 + 30 \div 3 \bullet 2$
D.	$5 \bullet 6 \div 3$	$3 \bullet 2^2 \div (6 - 3)$	$2 \bullet 3 + 10 \div 2$
E.	$29 + 21 - 25$	$36 \div 9 - 2$	$48 \div (6 \bullet 2)$
F.	$4 \bullet 3 + 2 - 7$	$45 \div 15 + 2 \bullet 3$	$3 + 7 \bullet 5 - 1$
G.	$3 \bullet (8 - 5)$	$(20 + 12) \div (4 + 4)$	$(15 - 3) \div 12 + 1$
H.	$4^2 - 5 \bullet 3$	$42 \div 7 + 2^2$	$5 \bullet (8 - 2) \bullet 3$
I.	$3^3 \div (2^3 + 1)$	$4 \bullet 5 - 3 \bullet 5 + 4$	$54 + 36 \div 3 - 30$

Name ______________________________ Translating algebraic expressions

What Do You Mean?

Use numbers and symbols to translate the expressions.

A. The sum of 8 and x is 15.

$8 + x = 15$

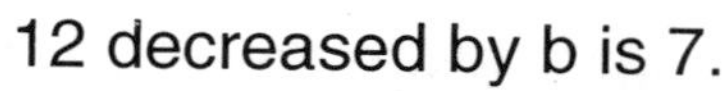

B. 7 more than n is 12. ______________

12 decreased by b is 7. ______________

C. 9 increased by c is 20. ______________

s decreased by 9 is 6. ______________

D. 6 more than 27 is s. ______________

8 taken away from t is 9. ______________

E. 8 more than v is 13. ______________

16 added to r is 87. ______________

F. The sum of x and 15 is 75. ______________

47 decreased by p is 12. ______________

G. 7 decreased by g is 1. ______________

y added to 14 is 35. ______________

H. h increased by 16 is 17. ______________

m decreased by 7 is 23. ______________

I. 23 increased by x is 94. ______________

110 decreased by 86 is x. ______________

J. p taken away from 20 is 6. ______________

26 less than j is 26. ______________

Number Patterns

Use the rule given to complete each function table.

A.

x	3	5	7	9	11
x + 2	5				

B.

e	1	5	9	13	17
e + 9					

g	2	4	6	8	10
10 – g					

C.

k	30	45	60	75	90
k – 15					

t	13	28	43	58	73
37 + t					

D.

w	91	82	73	64	55
100 – w					

f	11	22	33	44	55
f + 99					

Write the rule for each function table. Then complete the table.

E.

h	5	6	7	8	9
h + 3	8	9			

j	15	20	25	30	35
	16	21			

F.

m	10	12	14	16	18
	6	8			

p	28	35	42	49	56
	21	28			

G.

r	56	65	74	83	92
	46	55			

v	90	80	70	60	50
	79	69			

H.

y	37	35	33	31	29
	28	26			

b	59	62	65	68	71
	62	65			

Name ____________________

Find the Values

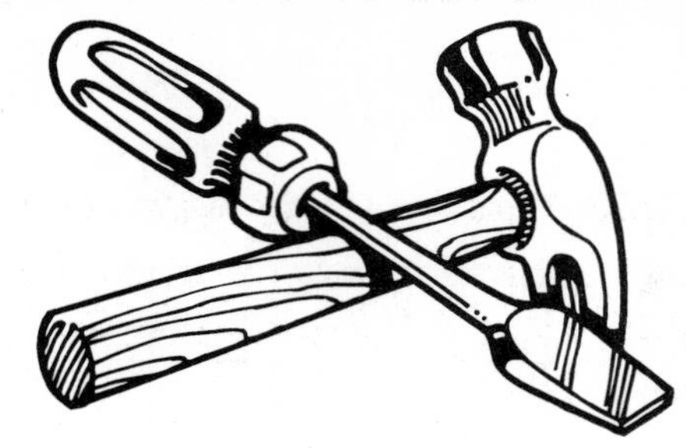

Evaluate each expression by replacing each variable (letter) with its given value.

Let x = 10.

A. $3.5 + x =$ ______ $12 - x =$ ______ $(8 + 9) - x =$ ______

B. $14 + x =$ ______ $43.2 - x =$ ______ $x - 5.3 =$ ______

Let y = 14.5.

C. $9 + y =$ ______ $56 - y =$ ______ $14.5 - y =$ ______

D. $53.5 + y =$ ______ $y - 9.3 =$ ______ $15 - y =$ ______

Let m = 25. Let n = 10.

E. $43 + (m + n) =$ ______ $50 + (m - n) =$ ______ $(95 + m) - n =$ ______

F. $(67 - m) + n =$ ______ $(m - 10) + n =$ ______ $n - 6 + m =$ ______

G. $49 - (m + n) =$ ______ $(m - n) - 15 =$ ______ $n + (30 - m) =$ ______

Let r = 9. Let s = 8. Let t = 5.

H. $(s + s) - 5 =$ ______ $(9 + 57) + (r - t) =$ ______ $(r - s) + t =$ ______

I. $t + (r - s) =$ ______ $(12 + t) - (r + s) =$ ______ $(r - s) + 9 =$ ______

J. $16 + (t + r) =$ ______ $100 + t - r =$ ______ $50 + (r - t) =$ ______

K. $(r + s) - 16 =$ ______ $87 - (r + s + t) =$ ______ $15 - (r + t) =$ ______

Replacement Sets

Solve each equation using a number from the given replacement set. If none of the numbers in the replacement set make the equation true, write **NS** (no solution).

Use the replacement set {0, 2, 4, 6}.

A. $51 - x = 45$ $x =$ 6 | $b + 79 = 81$ $b =$ ______ | $17 - a = 17$ $a =$ ______

B. $y - 15 = 82$ $y =$ ______ | $31 = 27 + f$ $f =$ ______ | $83 - 76 = g$ $g =$ ______

Use the replacement set {7, 8, 9, 10}.

C. $z + 8 = 17$ $z =$ ______ | $c - 15 = 7$ $c =$ ______ | $14 + h = 23$ $h =$ ______

D. $12 + n = 20$ $n =$ ______ | $p - 2 = 8$ $p =$ ______ | $r + 43 = 50$ $r =$ ______

Use the replacement set {25, 26, 27, 28, 29, 30}.

E. $35 + q = 60$ $q =$ ______ | $s + 29 = 68$ $s =$ ______ | $93 - i = 65$ $i =$ ______

F. $84 - 46 = j$ $j =$ ______ | $x - 13 = 17$ $x =$ ______ | $d - 47 = 17$ $d =$ ______

Use the replacement set {99, 100, 101}.

G. $47 + x = 100$ $x =$ ______ | $z - 65 = 35$ $z =$ ______ | $k - 2 = 99$ $k =$ ______

H. $m - 37 = 67$ $m =$ ______ | $v + 46 = 145$ $v =$ ______ | $176 - n = 75$ $n =$ ______

Name ______________________________

Addition Equations

$x + 9 = 28$
$x + 9 - \mathbf{9} = 28 - \mathbf{9}$
$x = 19$

To solve an addition equation, subtract the same number from both sides to make the variable stand alone.

A. $x + 8 = 13$ $4 + x = 12$ $x + 6 = 17$

B. $x + 5 = 15$ $x + 21 = 34$ $x + 38 = 81$

C. $x + 29 = 65$ $46 + x = 95$ $x + 6.6 = 7.2$

D. $x + 19 = 48$ $3.6 + x = 4.9$ $815 + x = 902$

E. $x + 13.8 = 15.6$ $347 + x = 409$ $x + 12 = 30.1$

F. $169 + x = 490$ $x + 6.7 = 24.5$ $196 + x = 500$

Name ____________________

Subtraction Equations

$y - 21 = 32$
$y - 21 + \mathbf{21} = 32 + \mathbf{21}$
$y = 53$

To solve a subtraction equation, add the same number to both sides to make the variable stand alone.

A. $y - 12 = 25$ $y - 19 = 21$ $y - 68 = 229$

B. $y - 56 = 7$ $y - 42 = 67$ $y - 35 = 16$

C. $y - 18 = 5$ $y - 4.8 = 9.2$ $y - 5.8 = 15.3$

D. $y - 18.7 = 4.2$ $y - 96 = 107$ $y - 62.5 = 83.1$

E. $y - 10.5 = 4.37$ $y - 73 = 196$ $y - 275 = 489$

F. $y - 14.6 = 9.8$ $y - 8.2 = 100$ $y - 96 = 1.4$

Name ____________________

Addition and Subtraction Equations

Solve the equations.

A. $n + 23 = 50$ | $x - 16 = 90$

B. $a + 47 = 85$ | $y + 2.9 = 9.2$ | $c + 0.76 = 1.54$

C. $n + 67 = 282$ | $y + 6.9 = 14.5$ | $x - 0.58 = 1.39$

D. $y - 77 = 229$ | $c - 167 = 85$ | $n - 25.8 = 19.7$

E. $c - 376 = 488$ | $x - 8.9 = 17.6$ | $n + 89 = 134$

F. $n + 38 = 84$ | $x + 5.6 = 9.2$ | $z - 6.85 = 4.76$

Fraction Solutions

Solve the equations.

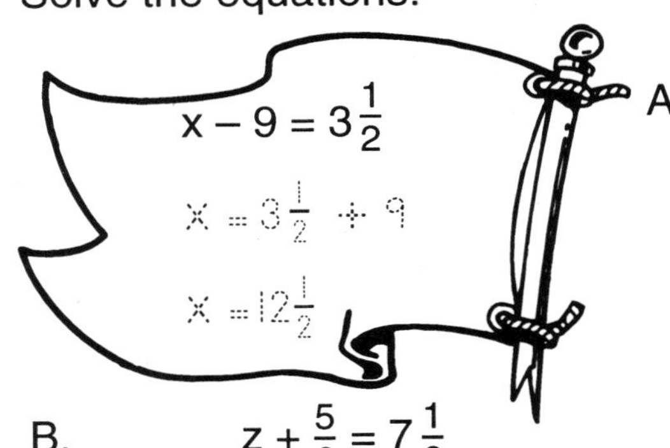

A. $n - 8\frac{1}{3} = 2\frac{2}{3}$ $\quad 1\frac{1}{4} + w = 4\frac{3}{4}$

B. $z + \frac{5}{6} = 7\frac{1}{6}$ $\quad d - 5\frac{1}{8} = 7\frac{1}{8}$ $\quad 8\frac{3}{4} + z = 10\frac{1}{4}$

C. $g - 4\frac{2}{3} = 8\frac{2}{3}$ $\quad y + \frac{1}{3} = \frac{2}{3}$ $\quad p - 10 = 3\frac{3}{4}$

D. $6\frac{1}{5} + n = 8\frac{2}{5}$ $\quad k - 5\frac{3}{4} = 9\frac{1}{4}$ $\quad \frac{11}{12} = w + \frac{5}{12}$

E. $a - 2\frac{2}{9} = 1\frac{1}{9}$ $\quad m - 4\frac{3}{10} = 1\frac{1}{10}$ $\quad n + 1\frac{2}{3} = 8$

F. $f + \frac{3}{14} = \frac{9}{14}$ $\quad f + 5\frac{3}{8} = 9\frac{1}{8}$ $\quad 9\frac{1}{3} - r = 2\frac{1}{3}$

Say It With Symbols

Translate each expression by writing an equation with numbers and a variable.

The product of 8 and n is 56. $8 \bullet n = 56$	A. The product of 12 and n is 132. ______
B. 60 divided by i is 15. ______	48 shared equally among 4 is z. ______
C. F divided by 5 is 5. ______	A number divided by 8 is 2. ______
D. Double a number r is 12. ______	35 is the product of m and 7. ______
E. The product of n and 8 is 108. ______	G divided by 25 is 8. ______
F. 82 shared equally among k is 20.5. ______	Triple a number d is 96. ______
G. The product of y and 7 is 84. ______	19 is the quotient of 76 divided by x. ______
H. 150 divided by w is 6. ______	B divided by 5 is 14. ______
I. 96 divided by n is 4. ______	8 times v is 168. ______
J. 46 is the quotient of p divided by 3. ______	78 is the product of 13 and q. ______

Name ____

More Number Patterns

Use the rules given to complete the function tables.

A.

x	5	6	7	8	9
3x	15				

y	10	12	14	16	18
$\frac{y}{2}$					

B.

z	5	10	15	20	25
6z					

b	4	8	12	16	20
$\frac{b}{4}$					

C.

c	20	21	22	23	24
20c					

f	10	20	30	40	50
$\frac{f}{5}$					

D.

h	10	15	20	25	30
$\frac{h}{10}$					

m	1	2	3	4	5
1.5m					

For each function table, write the rule and complete the table.

E.

g	1	2	3	4	5
3g	3	6			

i	5	10	15	20	25
	1	2			

F.

k	10	12	14	16	18
	5	6			

n	50	60	70	80	90
	10	12			

G.

r	3	5	7	9	11
	60	100			

p	38	57	76	95	114
	2	3			

Name ______________________________

Multiplication and Division Expressions

Evaluate each expression for the given values.

Let x = 15.

A. $3x =$ 45 $\frac{x}{5} =$ ______ $2x \div 3 =$ ______

B. $2.5x =$ ______ $105 \div x =$ ______ $\frac{2}{3}x =$ ______

Let y = 150. Let z = 32.

C. $\frac{z}{16} =$ ______ $\frac{y}{10} =$ ______ $\frac{2y}{3} =$ ______

D. $4y =$ ______ $y \div 30 =$ ______ $\frac{z}{0.5} =$ ______

Let a = 5. Let b = 15.

E. $3b \div 5 =$ ______ $\frac{18a}{3} =$ ______ $4b \bullet a =$ ______

F. $ab \div 3 =$ ______ $\frac{1}{3}b \div a =$ ______ $\frac{b}{a} =$ ______

G. $\frac{ab}{10} =$ ______ $\frac{30}{2a} =$ ______ $\frac{30}{2b} =$ ______

Let r = 200. Let s = 28. Let t = 10.

H. $\frac{r}{4} =$ ______ $3s =$ ______ $\frac{10s}{2} =$ ______

I. $\frac{r}{2t} =$ ______ $2s \bullet t =$ ______ $\frac{t}{r} =$ ______

J. $\frac{st}{2.8} =$ ______ $9t =$ ______ $\frac{s}{4} \bullet t =$ ______

Name ____________________ Using a replacement set

What Will Replace the Variable?

Solve each equation using a number from the given replacement set. If none of the numbers in the replacement set make the equation true, write **NS** (no solution).

Use the replacement set {0, 3, 6, 9, 12}.

A. $5 \bullet x = 30$ $x =$ 6 $b \bullet 12 = 0$ $b =$ ______ $72 \div g = 9$ $g =$ ______

B. $3 \div y = 1$ $y =$ ______ $z \div 4 = 3$ $z =$ ______ $7 \bullet z = 84$ $z =$ ______

Use the replacement set {10, 20, 30, 40}.

C. $6 \bullet f = 150$ $f =$ ______ $\frac{a}{6} = 5$ $a =$ ______ $9 \bullet g = 360$ $g =$ ______

D. $320 = 8 \bullet h$ $h =$ ______ $x \div 4 = 12$ $x =$ ______ $r \bullet 16 = 320$ $r =$ ______

Use the replacement set {1, 5, 25, 125}.

E. $3 \bullet y + 1 = 16$ $y =$ ______ $z \div 5 = 25$ $z =$ ______ $18 + 3 \bullet t = 33$ $t =$ ______

F. $f \div 15 = 15$ $f =$ ______ $(8 \bullet g) + 5 = 45$ $g =$ ______ $75 \div p = 15$ $p =$ ______

G. $6 + (4 \bullet e) = 26$ $e =$ ______ $\frac{m}{5} + 5 = 10$ $m =$ ______ $63 \div x = 63$ $x =$ ______

H. $(y - 5) \div 2 = 60$ $y =$ ______ $1 + (h \div 7) = 14$ $h =$ ______ $18 + (p \bullet 7) = 193$ $p =$ ______

Name________________________________

Multiplication Equations

To solve a multiplication equation, divide both sides by the same number to make the variable stand alone.

$$15x = 45$$
$$15x \div \mathbf{15} = 45 \div \mathbf{15}$$
$$x = 3$$

A. $6x = 42$ $7x = 112$ $8x = 72$ $246 = 6x$

B. $7x = 49$ $33x = 66$ $6x = 96$ $13x = 169$

C. $5x = 280$ $192 = 6x$ $7x = 105$ $144 = 24x$

D. $13x = 52$ $5.2x = 26$ $21.6 = 5.4x$ $9 = 2.25x$

E. $248 = 8x$ $15x = 240$ $4.8x = 36$ $56 = 3.5x$

F. $250 = 10x$ $93x = 186$ $1.2x = 14.4$ $200 = 2.5x$

Name ____________________

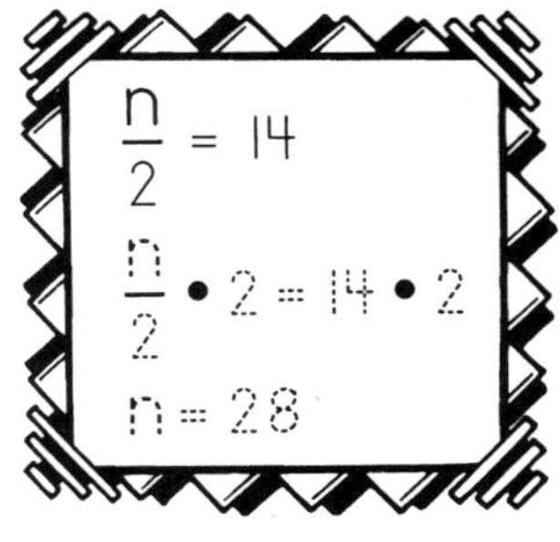

Division Equations

To solve a division equation, multiply both sides by the same number to make the variable stand alone.

A. $\frac{n}{7} = 21$ $\frac{n}{3} = 15$ $\frac{n}{4} = 10$ $\frac{n}{2} = 17$

B. $\frac{n}{6} = 72$ $\frac{n}{4} = 19$ $\frac{n}{5} = 25$ $\frac{n}{15} = 5$

C. $\frac{n}{8} = 25$ $\frac{n}{3} = 7.5$ $\frac{n}{15} = 14$ $\frac{n}{8} = 5.9$

D. $\frac{n}{3.5} = 12$ $\frac{n}{14} = 2.5$ $\frac{n}{10} = 8\frac{1}{2}$ $\frac{n}{4.7} = 93$

E. $\frac{n}{5} = 4.6$ $\frac{n}{2} = 3\frac{1}{3}$ $\frac{n}{9} = 27$ $\frac{n}{5.1} = 20.4$

F. $\frac{n}{10.9} = 2$ $\frac{n}{15.2} = 4.9$ $\frac{n}{6} = 42.7$ $\frac{n}{100} = \frac{2}{5}$

Name ______________________________

Multiplication and Division Equations

Solve the equations.

A. $8b = 72$ $240 = 8x$ $3 = \frac{p}{70}$

B. $4a = 28$ $10 = \frac{n}{3}$ $350w = 700$ $100 = 5z$

C. $\frac{x}{8} = 9$ $125e = 250$ $\frac{b}{10} = 8$ $250 = 2.5z$

D. $\frac{r}{1{,}000} = 7$ $24b = 312$ $\frac{n}{1.2} = 1.8$ $47n = 423$

E. $\frac{t}{27} = 36$ $5.27 = 3.1n$ $23 = \frac{c}{35}$ $\frac{b}{7.5} = 2.8$

F. $\frac{x}{2.5} = 0.25$ $\frac{d}{54} = 1.83$ $0.4c = 68$ $0.09n = 27$

Name ____

Understanding Integers

The set of integers contains all positive whole numbers and their negative opposites. Write an integer suggested by each situation listed below.

A.	a savings of $10	+10	a loss of 7 points	______
B.	a gain of 4 yards	______	5 miles below sea level	______
C.	a decrease of 15 pounds	______	10 seconds before liftoff	______
D.	3 feet under water	______	100 feet above sea level	______
E.	a 12-foot-deep crater	______	a 15° drop in temperature	______
F.	an expense of $39	______	a 20-yard penalty	______
G.	50 years ago	______	earnings of $45	______
H.	a profit of $150	______	14 years from now	______
I.	a debt of $175	______	a stock price drop of $1	______
J.	a 17° rise in temperature	______	6 laps behind the lead car	______
K.	a $25 profit	______	a $50 bonus	______

Name____

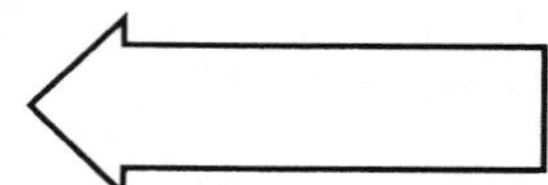

Nifty Number Lines

Give the integer for each point on the number lines.

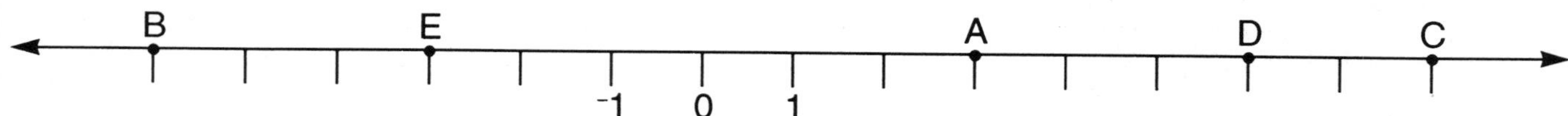

A = ________ B = ________ C = ________ D = ________ E = ________

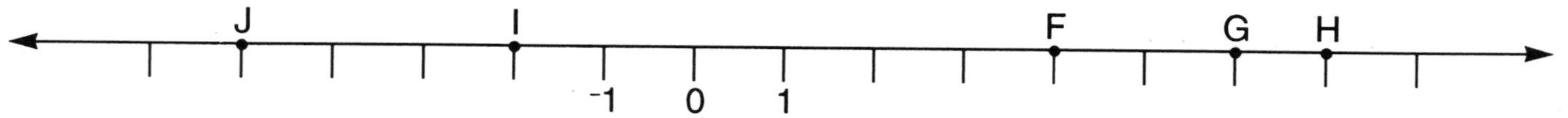

F = ________ G = ________ H = ________ I = ________ J = ________

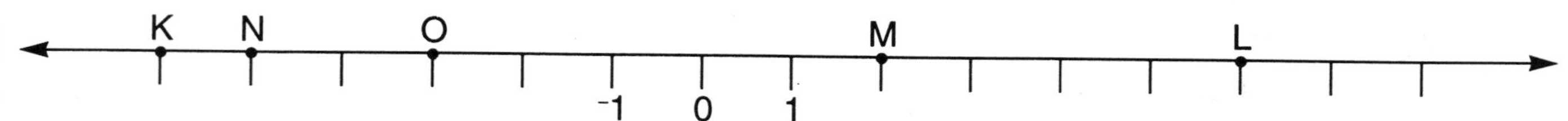

K = ________ L = ________ M = ________ N = ________ O = ________

Arrange the numbers on the number lines from the least to the greatest.

P. 0, 1, 6, -2, -5

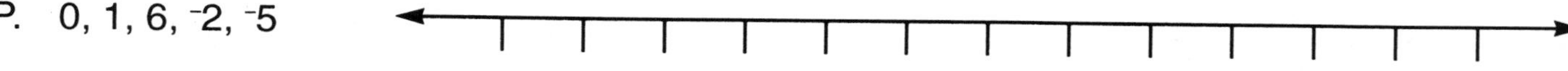

Q. 1, -3, 7, 2, -4

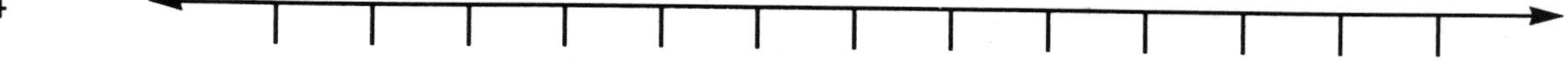

R. -3, -2, 1, -8, 3

S. 3, 4, -5, 5, -1

T. 0, -6, -7, 3, 4,

U. -7, -4, -1, -9, 2

Name ____________________

Comparing Integers Using Absolute Value

The absolute value of an integer is its distance from zero on the number line. Bars around a number indicate absolute value.

> $|5| = 5$ means the absolute value of 5 is 5.
> $|-5| = 5$ means the absolute value of -5 is 5.

Compare each pair of numbers. Write <, >, or = in the circle.

A.	$|-21| \bigcirc 1$	$|3| \bigcirc |-3|$	$0 \bigcirc -1$	$5 \bigcirc |4|$
B.	$|-3| \bigcirc |2|$	$5 \bigcirc |-6|$	$|-9| \bigcirc |9|$	$|-1| \bigcirc |1|$
C.	$|-9| \bigcirc 8$	$12 \bigcirc |-1|$	$|-4| \bigcirc 0$	$5 \bigcirc |-5|$
D.	$|-6| \bigcirc |0|$	$9 \bigcirc |-4|$	$5 \bigcirc |-6|$	$8 \bigcirc |-8|$
E.	$|-7| \bigcirc 8$	$|-9| \bigcirc 10$	$|-7| \bigcirc 8$	$|-9| \bigcirc |-10|$
F.	$|12| \bigcirc -1$	$|-12| \bigcirc 2$	$-5 \bigcirc |-6|$	$7 \bigcirc |-9|$
G.	$-4 \bigcirc 4$	$|-1| \bigcirc |-2|$	$15 \bigcirc |-14|$	$|16| \bigcirc |-16|$
H.	$-2 \bigcirc |3|$	$|-14| \bigcirc |14|$	$-17 \bigcirc -16$	$|-23| \bigcirc |21|$
I.	$-4 \bigcirc -5$	$4 \bigcirc |-5|$	$|-23| \bigcirc |-25|$	$14 \bigcirc |-15|$
J.	$-16 \bigcirc -18$	$|-18| \bigcirc |-17|$	$|-8| \bigcirc |8|$	$-12 \bigcirc |-12|$
K.	$|-11| \bigcirc |-13|$	$0 \bigcirc |-3|$	$0 \bigcirc |-2|$	$9 \bigcirc |-10|$

Adding Integers

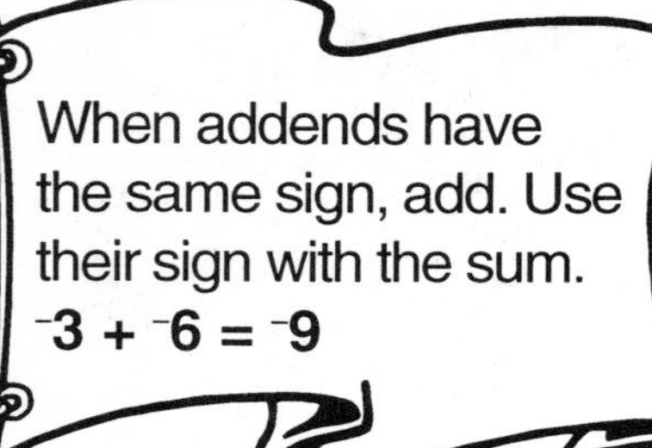
When addends have
the same sign, add. Use
their sign with the sum.
${}^-3 + {}^-6 = {}^-9$

When addends have different signs, subtract. Use the sign of the greater addend.
$5 + {}^-8 = {}^-3$

A.	$9 + {}^-4 =$ ____	$5 + {}^-7 =$ ____	${}^-3 + {}^-7 =$ ____
B.	${}^-7 + 2 =$ ____	$10 + {}^-8 =$ ____	${}^-8 + 4 =$ ____
C.	$9 + {}^-3 =$ ____	${}^-3 + {}^-6 =$ ____	${}^-3 + 11 =$ ____
D.	${}^-8 + 14 =$ ____	${}^-5 + 13 =$ ____	${}^-9 + 25 =$ ____
E.	$3 + {}^-12 =$ ____	$17 + {}^-8 =$ ____	${}^-24 + 0 =$ ____
F.	${}^-26 + {}^-23 =$ ____	${}^-8 + {}^-5 =$ ____	$5 + {}^-14 =$ ____
G.	${}^-14 + {}^-16 =$ ____	$35 + 29 =$ ____	${}^-34 + {}^-26 =$ ____
H.	${}^-4 + 6 =$ ____	$25 + {}^-9 =$ ____	$17 + 18 =$ ____
I.	${}^-7 + {}^-3 =$ ____	${}^-9 + {}^-1 =$ ____	${}^-4 + 5 =$ ____
J.	$3 + {}^-9 =$ ____	${}^-6 + 10 =$ ____	$9 + {}^-9 =$ ____
K.	${}^-8 + 9 =$ ____	$23 + {}^-5 =$ ____	${}^-24 + 25 =$ ____
L.	${}^-7 + 3 =$ ____	$10 + {}^-19 =$ ____	$26 + {}^-28 =$ ____

Name ______________________________

Subtracting Integers

To subtract an integer, add its opposite.

$4 - 7$ becomes $4 + {}^{-}7 = {}^{-}3$

$3 - {}^{-}9$ becomes $3 + 9 = 12$

A.	$15 - 7 =$ ______	$0 - {}^{-}6 =$ ______	$4 - 9 =$ ______
B.	$0 - 7 =$ ______	$14 - {}^{-}5 =$ ______	$8 - {}^{-}6 =$ ______
C.	$6 - {}^{-}9 =$ ______	$12 - {}^{-}12 =$ ______	$9 - {}^{-}15 =$ ______
D.	${}^{-}15 - {}^{-}9 =$ ______	$8 - 12 =$ ______	${}^{-}15 - 9 =$ ______
E.	${}^{-}15 - {}^{-}2 =$ ______	$5 - 5 =$ ______	${}^{-}8 - {}^{-}8 =$ ______
F.	${}^{-}11 - {}^{-}5 =$ ______	${}^{-}13 - {}^{-}5 =$ ______	${}^{-}6 - {}^{-}8 =$ ______
G.	${}^{-}2 - {}^{-}6 =$ ______	$6 - {}^{-}15 =$ ______	${}^{-}20 - 6 =$ ______
H.	$17 - {}^{-}8 =$ ______	${}^{-}4 - 25 =$ ______	${}^{-}4 - {}^{-}25 =$ ______
I.	$4 - 21 =$ ______	$5 - {}^{-}15 =$ ______	${}^{-}5 - 15 =$ ______
J.	$9 - 16 =$ ______	${}^{-}14 - {}^{-}12 =$ ______	$14 - {}^{-}12 =$ ______
K.	${}^{-}10 - 11 =$ ______	${}^{-}5 - {}^{-}25 =$ ______	${}^{-}9 - 43 =$ ______
L.	${}^{-}14 - 12 =$ ______	${}^{-}10 - {}^{-}11 =$ ______	$10 - {}^{-}11 =$ ______

Name ______________________________

Integer Sums and Differences

Add or subtract the integers.

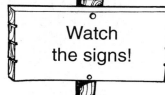

Watch the signs!

A.	$2 + {}^{-}5 =$ ______	$2 - {}^{-}5 =$ ______	
B.	$3 + {}^{-}10 =$ ______	${}^{-}3 - 10 =$ ______	$15 - 6 =$ ______
C.	${}^{-}15 + 6 =$ ______	${}^{-}15 - {}^{-}6 =$ ______	$20 - 10 =$ ______
D.	$20 - {}^{-}10 =$ ______	$17 - 27 =$ ______	${}^{-}17 - 27 =$ ______
E.	$14 - 5 =$ ______	$20 + {}^{-}10 =$ ______	${}^{-}14 - {}^{-}5 =$ ______
F.	$29 - 5 =$ ______	$14 - {}^{-}5 =$ ______	${}^{-}27 - {}^{-}17 =$ ______
G.	${}^{-}29 - 5 =$ ______	$19 - {}^{-}9 =$ ______	${}^{-}19 + 9 =$ ______
H.	${}^{-}14 + {}^{-}5 =$ ______	$5 - 25 =$ ______	$26 + {}^{-}6 =$ ______
I.	${}^{-}19 - 9 =$ ______	$19 + {}^{-}9 =$ ______	$49 - 49 =$ ______
J.	${}^{-}26 - 6 =$ ______	${}^{-}49 + 49 =$ ______	${}^{-}26 - {}^{-}6 =$ ______
K.	$17 - 46 =$ ______	$19 - 9 =$ ______	${}^{-}96 - {}^{-}47 =$ ______
L.	${}^{-}53 + {}^{-}53 =$ ______	$96 - 47 =$ ______	${}^{-}83 + 82 =$ ______
M.	${}^{-}53 - {}^{-}53 =$ ______	$46 - 17 =$ ______	${}^{-}83 - {}^{-}82 =$ ______

Name ____

Multiplying Integers

The product is positive if both factors are positive or if both are negative. The product is negative if one factor is positive and one is negative.

A.	$4 \cdot 5 =$ ______	$^{-}4 \cdot 9 =$ ______	$8 \cdot {^{-}6} =$ ______
B.	$^{-}8 \cdot {^{-}9} =$ ______	$^{-}5 \cdot 7 =$ ______	$2 \cdot {^{-}3} =$ ______
C.	$^{-}4 \cdot 11 =$ ______	$^{-}2 \cdot {^{-}13} =$ ______	$12 \cdot 12 =$ ______
D.	$^{-}9 \cdot {^{-}7} =$ ______	$^{-}8 \cdot 5 =$ ______	$^{-}12 \cdot 6 =$ ______
E.	$^{-}6 \cdot {^{-}8} =$ ______	$29 \cdot {^{-}6} =$ ______	$5 \cdot 7 =$ ______
F.	$^{-}11 \cdot 10 =$ ______	$^{-}10 \cdot {^{-}4} =$ ______	$15 \cdot {^{-}8} =$ ______
G.	$^{-}9 \cdot 11 =$ ______	$6 \cdot {^{-}7} =$ ______	$^{-}4 \cdot {^{-}6} =$ ______
H.	$8 \cdot 7 =$ ______	$14 \cdot {^{-}9} =$ ______	$^{-}8 \cdot 16 =$ ______
I.	$^{-}9 \cdot 8 =$ ______	$12 \cdot {^{-}11} =$ ______	$^{-}5 \cdot {^{-}3} =$ ______
J.	$^{-}5 \cdot {^{-}11} =$ ______	$^{-}8 \cdot 12 =$ ______	$^{-}2 \cdot {^{-}7} =$ ______
K.	$^{-}13 \cdot {^{-}6} =$ ______	$^{-}9 \cdot 5 =$ ______	$^{-}12 \cdot {^{-}3} =$ ______
L.	$4 \cdot {^{-}12} =$ ______	$9 \cdot 9 =$ ______	$^{-}11 \cdot {^{-}11} =$ ______
M.	$5 \cdot {^{-}13} =$ ______	$15 \cdot {^{-}7} =$ ______	$^{-}6 \cdot {^{-}9} =$ ______
N.	$^{-}7 \cdot {^{-}15} =$ ______	$^{-}14 \cdot 5 =$ ______	$9 \cdot {^{-}13} =$ ______

Dividing Integers

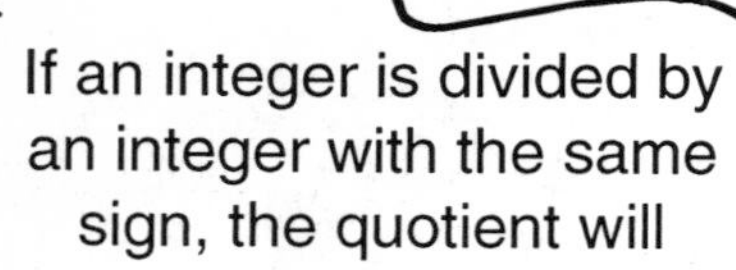

If an integer is divided by an integer with the same sign, the quotient will be positive.

$^{-}72 \div {}^{-}8 = 9$

If an integer is divided by an integer with the opposite sign, the quotient will be negative.

$48 \div {}^{-}6 = {}^{-}8$

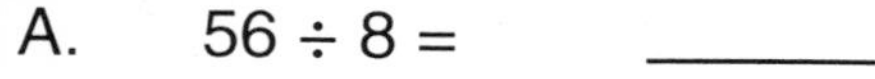

A.	$56 \div 8 =$ ______	$^{-}24 \div {}^{-}8 =$ ______	$^{-}20 \div {}^{-}4 =$ ______
B.	$81 \div 9 =$ ______	$^{-}36 \div 4 =$ ______	$42 \div {}^{-}7 =$ ______
C.	$^{-}64 \div 8 =$ ______	$72 \div {}^{-}9 =$ ______	$45 \div {}^{-}5 =$ ______
D.	$36 \div 6 =$ ______	$^{-}48 \div 8 =$ ______	$66 \div {}^{-}11 =$ ______
E.	$52 \div {}^{-}4 =$ ______	$^{-}90 \div 10 =$ ______	$9 \div {}^{-}1 =$ ______
F.	$60 \div {}^{-}12 =$ ______	$^{-}36 \div {}^{-}3 =$ ______	$96 \div {}^{-}3 =$ ______
G.	$30 \div {}^{-}5 =$ ______	$144 \div 12 =$ ______	$^{-}72 \div 6 =$ ______
H.	$^{-}81 \div {}^{-}9 =$ ______	$40 \div {}^{-}8 =$ ______	$105 \div {}^{-}15 =$ ______
I.	$28 \div 7 =$ ______	$25 \div {}^{-}5 =$ ______	$^{-}144 \div 4 =$ ______
J.	$15 \div {}^{-}3 =$ ______	$^{-}100 \div {}^{-}25 =$ ______	$^{-}16 \div {}^{-}16 =$ ______
K.	$56 \div {}^{-}7 =$ ______	$98 \div {}^{-}7 =$ ______	$^{-}14 \div 2 =$ ______
L.	$^{-}121 \div {}^{-}11 =$ ______	$100 \div {}^{-}100 =$ ______	$65 \div {}^{-}5 =$ ______
M.	$76 \div {}^{-}2 =$ ______	$^{-}110 \div 5 =$ ______	$80 \div {}^{-}4 =$ ______

Name ____

Integer Products and Quotients

Watch the signs!

Multiply or divide the integers.

A.	$5 \bullet {}^{-}10 =$ ______	${}^{-}16 \div 2 =$ ______	
B.	${}^{-}7 \bullet {}^{-}9 =$ ______	${}^{-}12 \div 3 =$ ______	${}^{-}12 \bullet {}^{-}9 =$ ______
C.	${}^{-}8 \div {}^{-}2 =$ ______	${}^{-}15 \div {}^{-}3 =$ ______	$8 \bullet {}^{-}8 =$ ______
D.	${}^{-}16 \bullet {}^{-}6 =$ ______	${}^{-}75 \div {}^{-}5 =$ ______	${}^{-}48 \div 6 =$ ______
E.	${}^{-}80 \div 5 =$ ______	$66 \div {}^{-}11 =$ ______	$5 \bullet {}^{-}12 =$ ______
F.	${}^{-}11 \bullet {}^{-}11 =$ ______	$10 \bullet {}^{-}9 =$ ______	${}^{-}12 \bullet 5 =$ ______
G.	${}^{-}81 \div 9 =$ ______	$9 \bullet 11 =$ ______	${}^{-}90 \div 5 =$ ______
H.	$45 \div {}^{-}9 =$ ______	$12 \bullet {}^{-}9 =$ ______	${}^{-}13 \bullet 3 =$ ______
I.	${}^{-}15 \bullet {}^{-}6 =$ ______	${}^{-}12 \bullet {}^{-}8 =$ ______	${}^{-}96 \div 8 =$ ______
J.	${}^{-}5 \bullet {}^{-}8 =$ ______	${}^{-}72 \div 9 =$ ______	${}^{-}100 \div {}^{-}10 =$ ______
K.	${}^{-}11 \div 11 =$ ______	${}^{-}8 \bullet {}^{-}8 =$ ______	${}^{-}12 \div 1 =$ ______
L.	$28 \div 7 =$ ______	$42 \div {}^{-}6 =$ ______	$25 \bullet {}^{-}5 =$ ______
M.	$12 \div {}^{-}1 =$ ______	$11 \div {}^{-}11 =$ ______	$4 \bullet 12 =$ ______
N.	${}^{-}12 \div {}^{-}12 =$ ______	$108 \div {}^{-}12 =$ ______	$11 \bullet {}^{-}12 =$ ______

Name ____

Evaluating Expressions With Integers

Evaluate each expression for the given values.

Let r = $^{-}4$.

A.	$5r =$ $^{-}20$	$r + 5 =$ ______	$r - 12 =$ ______
B.	$\frac{20}{r} =$ ______	$2.5r =$ ______	$^{-}6r =$ ______

Let a = $^{-}8$. Let b = 6.

C.	$a + b =$ ______	$a - b =$ ______	$ab =$ ______
D.	$a - 10 =$ ______	$^{-}2ab =$ ______	$\frac{4b}{a} =$ ______

Let x = $^{-}3$. Let y = $^{-}12$.

E.	$y + 10 =$ ______	$^{-}3x =$ ______	$\frac{y}{3} =$ ______
F.	$y - x =$ ______	$xy =$ ______	$xy \div {}^{-}4 =$ ______
G.	$15x =$ ______	$\frac{y}{2x} =$ ______	$14 + y =$ ______

Let r = $^{-}4$. Let s = $^{-}1$. Let t = $^{-}16$.

H.	$^{-}4r =$ ______	$4rs =$ ______	$t \div r =$ ______
I.	$t + 20 =$ ______	$12 - t =$ ______	$15s =$ ______
J.	$\frac{t}{4rs} =$ ______	$5r =$ ______	$\frac{r}{s} =$ ______

Name ______________________________

Integer Solutions

Rewrite each equation so that the variable stands alone on one side. Then solve the equation.

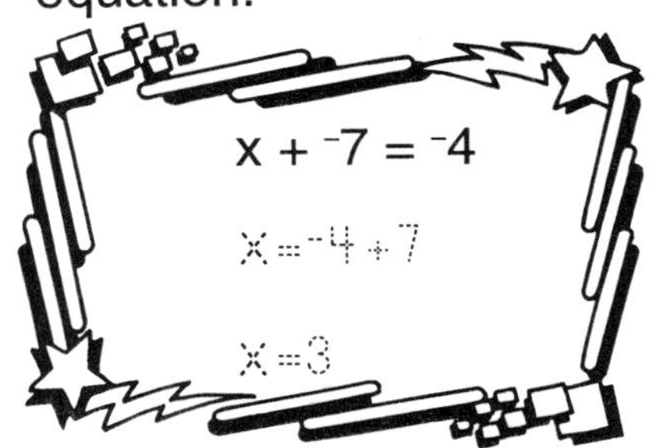

A. $y - 6 = {}^{-}18$ $\quad$ $6y = {}^{-}18$ $\quad$ $t + {}^{-}9 = 5$

B. ${}^{-}4s = 36$ $\quad$ ${}^{-}7c = {}^{-}56$ $\quad$ $n - {}^{-}2 = 8$ $\quad$ $y - 7 = {}^{-}3$

C. ${}^{-}5n = {}^{-}80$ $\quad$ $x - {}^{-}14 = 0$ $\quad$ $16 = n - 3$ $\quad$ $\frac{g}{7} = {}^{-}21$

D. $x - {}^{-}4 = 5$ $\quad$ $y \div {}^{-}11 = {}^{-}11$ $\quad$ $72 = {}^{-}9h$ $\quad$ $2t = {}^{-}64$

E. $16 + p = 4$ $\quad$ ${}^{-}30 = 5w$ $\quad$ $\frac{w}{{}^{-}3} = {}^{-}5$ $\quad$ ${}^{-}8y = {}^{-}64$

F. $\frac{c}{9} = {}^{-}2$ $\quad$ $n - {}^{-}18 = 3$ $\quad$ $z + 8 = {}^{-}5$ $\quad$ $m \div 14 = {}^{-}3$

Name ____

Graphing Ordered Pairs

Numbers in an ordered pair are used to locate a point on the coordinate plane. The first number (x-coordinate) tells you how far to move left or right. The second number (y-coordinate) tells you how far to move up or down.

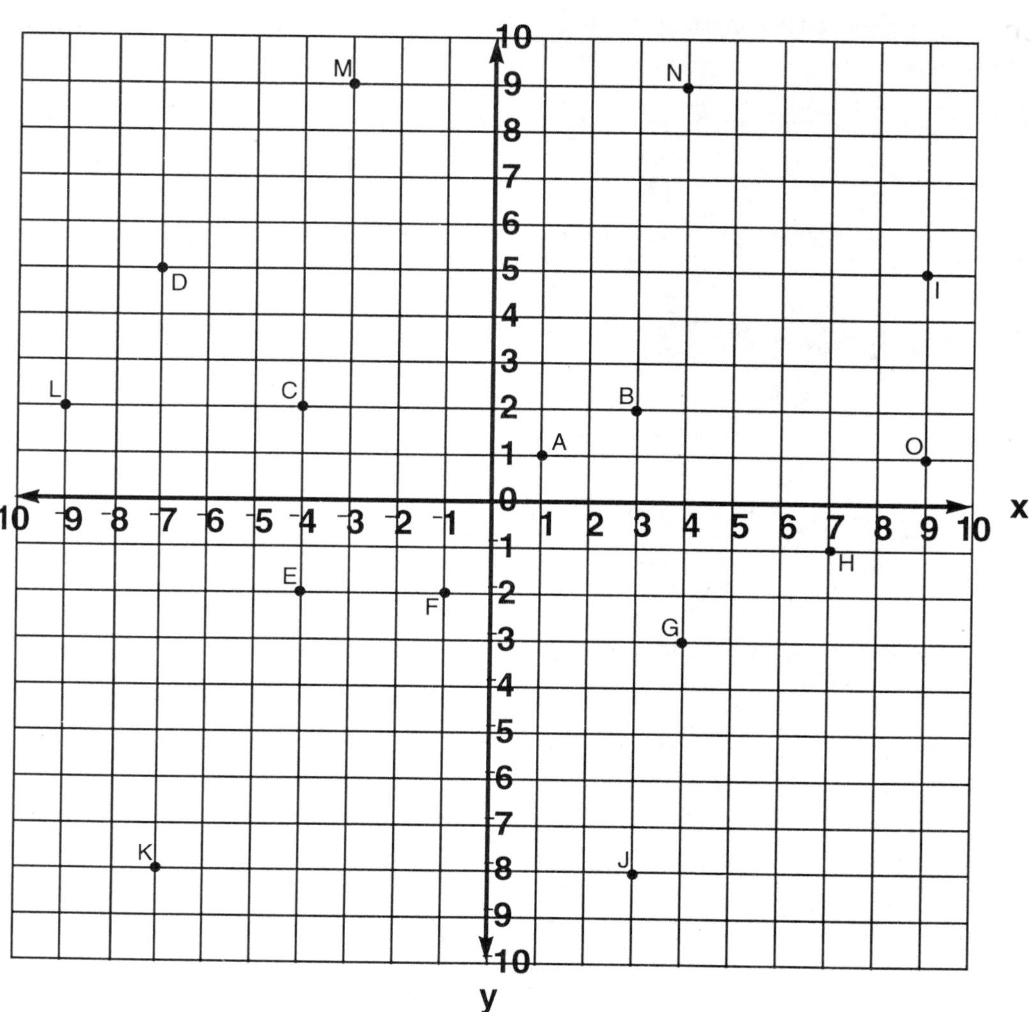

Give the ordered pair for each point named.

A ________ B ________ C ________ D ________ E ________

F ________ G ________ H ________ I ________ J ________

K ________ L ________ M ________ N ________ O ________

Draw and label each point at the given location.

P (5, 2) R (-5, 5) S (-9, -4) T (9, -4) U (8, -10)

V (3, -2) W (7, 7) X (-2, -2) Y (2, -7) Z (1, -9)

Name ______________________________

The Decimal Stops Here

Change each fraction to a decimal. Your answers will be **terminating decimals** because they end with a remainder of zero. You may use a calculator to check your work.

Remember to keep
dividing until
the remainder
is zero.

A. $\frac{3}{4}$ = 0.75 $\frac{1}{2}$ = ______ $\frac{1}{20}$ = ______

B. $\frac{5}{2}$ = ______ $\frac{17}{50}$ = ______ $\frac{4}{5}$ = ______

C. $\frac{9}{10}$ = ______ $\frac{19}{200}$ = ______ $\frac{1}{16}$ = ______ $\frac{19}{40}$ = ______

D. $\frac{7}{4}$ = ______ $\frac{8}{5}$ = ______ $\frac{4}{25}$ = ______ $\frac{3}{16}$ = ______

E. $\frac{13}{25}$ = ______ $\frac{3}{75}$ = ______ $\frac{13}{40}$ = ______ $\frac{1}{125}$ = ______

F. $\frac{3}{40}$ = ______ $\frac{9}{8}$ = ______ $\frac{5}{16}$ = ______ $\frac{9}{4}$ = ______

G. $\frac{31}{50}$ = ______ $\frac{14}{25}$ = ______ $\frac{19}{20}$ = ______ $\frac{11}{16}$ = ______

H. $\frac{21}{20}$ = ______ $\frac{129}{200}$ = ______ $\frac{91}{50}$ = ______ $\frac{15}{16}$ = ______

I. $\frac{41}{25}$ = ______ $\frac{26}{25}$ = ______ $\frac{17}{200}$ = ______ $\frac{13}{16}$ = ______

J. $\frac{3}{5}$ = ______ $\frac{8}{25}$ = ______ $\frac{21}{16}$ = ______ $\frac{11}{2}$ = ______

K. $\frac{51}{50}$ = ______ $\frac{127}{200}$ = ______ $\frac{19}{25}$ = ______ $\frac{11}{5}$ = ______

Name ____________________

Repeat After Me...

Repeating decimals do not terminate. You can draw a bar over the digit or digits that repeat. Rewrite each repeating decimal with a bar to show the repeating digit or digits.

0.1444 means the same as $0.1\overline{4}$.

A. 0.16666 = $0.1\overline{6}$ 0.3333 = ______

B. 0.161616 = ______   0.263263263 = ______

C. 0.161161161 = ______   0.113113113 = ______

D. 0.83333 = ______   0.416666 = ______

Rewrite each fraction as a decimal. Use a calculator to help you. Show the repeating digit or digits with a bar.

E.	$\frac{7}{9}$ = $0.\overline{7}$	$\frac{31}{15}$ = ______	$\frac{7}{3}$ = ______	$\frac{1}{12}$ = ______
F.	$\frac{2}{15}$ = ______	$\frac{19}{12}$ = ______	$\frac{11}{15}$ = ______	$\frac{8}{3}$ = ______
G.	$\frac{10}{11}$ = ______	$\frac{11}{6}$ = ______	$\frac{5}{18}$ = ______	$\frac{11}{12}$ = ______
H.	$\frac{7}{60}$ = ______	$\frac{26}{9}$ = ______	$\frac{20}{9}$ = ______	$\frac{3}{11}$ = ______
I.	$\frac{22}{3}$ = ______	$\frac{13}{36}$ = ______	$\frac{1}{60}$ = ______	$\frac{41}{3}$ = ______
J.	$\frac{7}{6}$ = ______	$\frac{4}{3}$ = ______	$\frac{4}{9}$ = ______	$\frac{1}{30}$ = ______
K.	$\frac{29}{3}$ = ______	$\frac{5}{9}$ = ______	$\frac{10}{3}$ = ______	$\frac{10}{9}$ = ______

Name ______________________________

Decimal Definitions

Use a calculator to change each fraction to a decimal. Circle **T** or **R** to identify each decimal as **terminating** or **repeating**.

A. $\frac{23}{40} =$ 0.575 (T) R

B. $\frac{9}{24} =$ ________ T R

C. $\frac{16}{25} =$ ________ T R

D. $\frac{88}{33} =$ ________ T R

E. $\frac{65}{18} =$ ________ T R

F. $\frac{17}{16} =$ ________ T R

G. $\frac{3}{11} =$ ________ T R

H. $\frac{2}{9} =$ ________ T R

I. $\frac{15}{33} =$ ________ T R

J. $\frac{33}{16} =$ ________ T R

K. $\frac{49}{80} =$ ________ T R

L. $\frac{8}{9} =$ ________ T R

$\frac{11}{3} =$ ________ T R

$\frac{6}{11} =$ ________ T R

$\frac{48}{11} =$ ________ T R

$\frac{72}{99} =$ ________ T R

$\frac{44}{54} =$ ________ T R

$\frac{13}{11} =$ ________ T R

$\frac{47}{50} =$ ________ T R

$\frac{1}{6} =$ ________ T R

$\frac{7}{8} =$ ________ T R

$\frac{17}{18} =$ ________ T R

Name ______________________________

It's the Greatest!

Compare each pair of numbers. Write <, >, or = .
You may want to use a number line.

A.	$1.5 \bigcirc 1\frac{1}{2}$	$5\frac{3}{4} \bigcirc 5.8$	
B.	$\frac{4}{5} \bigcirc 0.45$	$2 \bigcirc \frac{3}{2}$	$\frac{9}{16} \bigcirc \frac{7}{12}$
C.	$3\frac{7}{20} \bigcirc 3.14$	$0.25 \bigcirc \frac{1}{5}$	$3\frac{5}{8} \bigcirc 3.6$
D.	$5\frac{3}{25} \bigcirc 5.15$	$\frac{11}{50} \bigcirc \frac{6}{25}$	$3\frac{4}{25} \bigcirc 3.16$
E.	$8\frac{3}{5} \bigcirc 8.6$	$\frac{3}{16} \bigcirc \frac{1}{6}$	$\frac{1}{9} \bigcirc 0.11$
F.	$4\frac{9}{20} \bigcirc 4.5$	$4.6 \bigcirc \frac{35}{8}$	$9\frac{2}{3} \bigcirc 9.6$
G.	$5\frac{1}{8} \bigcirc 5.2$	$8.3 \bigcirc 8\frac{1}{3}$	$2.75 \bigcirc 2\frac{3}{4}$
H.	$10\frac{1}{2} \bigcirc 10.12$	$4.9 \bigcirc 4\frac{9}{10}$	$3.21 \bigcirc 3\frac{11}{50}$
I.	$8\frac{1}{11} \bigcirc 8\frac{1}{12}$	$7.5 \bigcirc 7\frac{5}{9}$	$7\frac{3}{11} \bigcirc 7\frac{2}{10}$
J.	$5\frac{1}{12} \bigcirc 5.09$	$1.325 \bigcirc 1\frac{3}{8}$	$7\frac{9}{11} \bigcirc 7.8$
K.	$4\frac{1}{5} \bigcirc 4.2$	$3.87 \bigcirc 3\frac{7}{8}$	$5\frac{1}{33} \bigcirc 5.03$
L.	$8\frac{8}{11} \bigcirc 8.8$	$6.7 \bigcirc 6\frac{6}{7}$	$9\frac{9}{10} \bigcirc 9\frac{91}{100}$

Name ____

May I Take Your Order?

Write the numbers in order from the least to the greatest.

A. $^{-}5.8, {}^{-}7, 3\frac{5}{8}, 1\frac{2}{3}$

$\frac{7}{9}, \frac{^{-}5}{12}, \frac{^{-}7}{36}, {}^{-}0.5, 1.1$

B. $^{-}5, 10, 0.8, \frac{^{-}9}{10}, {}^{-}12$

$12\frac{3}{4}, {}^{-}16, 0, \frac{15}{16}, \frac{^{-}3}{4}$

C. $^{-}2\frac{1}{2}, 1, {}^{-}2, 1\frac{1}{2}, 0.5$

$\frac{3}{5}, 0.45, \frac{^{-}5}{10}, 0.4$

D. $1\frac{1}{2}, 2\frac{1}{2}, 1.4, {}^{-}1.55, 2\frac{1}{4}$

$\frac{^{-}1}{3}, {}^{-}0.6, 0.3, 0.03, {}^{-}0.03$

E. $1.45, 1\frac{2}{5}, {}^{-}2.7, {}^{-}2\frac{3}{4}, 1$

$\frac{^{-}3}{2}, \frac{^{-}4}{3}, {}^{-}0.5, 0.001, 0$

F. $^{-}1.75, 0.3, 1.5, \frac{^{-}1}{4}, \frac{^{-}8}{10}$

$\frac{9}{10}, {}^{-}0.9, {}^{-}1, 0.03, {}^{-}0.05$

Name ____________________

Integer Powers of Ten

Rewrite each number as a whole number or a fraction. If the exponent is negative, write a fraction with 1 as the numerator and the number with the exponent in positive form as the denominator. Then find the value of the denominator. You may use a calculator.

$5^5 = 3{,}125$ $\quad$ $8^{-2} = \frac{1}{8^2} = \frac{1}{64}$

A. $3^{-4} =$ ______ $\quad 2^{-6} =$ ______ $\quad 5^4 =$ ______

B. $2^{-4} =$ ______ $\quad 9^2 =$ ______ $\quad 4^{-3} =$ ______

C. $7^{-3} =$ ______ $\quad 10^{-4} =$ ______ $\quad 8^{-3} =$ ______

D. $2^6 =$ ______ $\quad 3^3 =$ ______ $\quad 4^{-4} =$ ______

E. $7^{-2} =$ ______ $\quad 2^{-7} =$ ______ $\quad 10^{-2} =$ ______

Write each expression as a number with a positive or negative exponent. If the number is a fraction, it will have a negative exponent.

$343 = 7^3$ $\quad$ $\frac{1}{16} = 2^{-4}$

F. $\frac{1}{81} =$ ______ $\quad \frac{1}{8} =$ ______ $\quad \frac{1}{9} =$ ______

G. $\frac{1}{64} =$ ______ $\quad \frac{1}{25} =$ ______ $\quad 27 =$ ______

H. $\frac{1}{100} =$ ______ $\quad 1{,}000 =$ ______ $\quad \frac{1}{121} =$ ______

I. $\frac{1}{216} =$ ______ $\quad 125 =$ ______ $\quad \frac{1}{169} =$ ______

J. $\frac{1}{10{,}000} =$ ______ $\quad \frac{1}{343} =$ ______ $\quad \frac{1}{256} =$ ______

Scientific Notation With Integer Exponents

A number is in scientific notation if it is written as the product of a number from 1 to 9 and a power of 10. Write each number using scientific notation. If the standard form number given is less than 1, the exponent will be negative.

A.	0.93	0.0004	0.0056	0.091
	______	______	______	______
B.	0.000045	0.0089	158	0.00306
	______	______	______	______
C.	0.07345	0.3	0.896	34,967
	______	______	______	______

Write the numbers in standard form. Watch the signs on the exponents!

D.	9×10^{-2}	4×10^{-4}	8.3×10^{-2}	1.95×10^{-3}
	0.09	______	______	______
E.	9.2×10^{3}	2.3×10^{-1}	6.03×10^{-2}	8.97×10^{-4}
	______	______	______	______
F.	7.25×10^{-4}	8.1×10^{-3}	4.083×10^{4}	6.98×10^{-3}
	______	______	______	______
G.	5.5×10^{-5}	4.78×10^{-1}	5.834×10^{3}	1.3×10^{-2}
	______	______	______	______

Square Roots

To find the square root of a number, find the number that when multiplied by itself is equal to the number. Every positive number has a positive square root and a negative square root.

A. $\sqrt{144} =$ ______ $\sqrt{16} =$ ______ $\sqrt{49} =$ ______

B. $-\sqrt{25} =$ ______ $\sqrt{\frac{1}{4}} =$ ______ $\sqrt{169} =$ ______

C. $\sqrt{64} =$ ______ $\sqrt{100} =$ ______ $-\sqrt{121} =$ ______ $\sqrt{81} =$ ______

D. $\sqrt{\frac{1}{16}} =$ ______ $-\sqrt{400} =$ ______ $\sqrt{\frac{9}{25}} =$ ______ $\sqrt{121} =$ ______

E. $\sqrt{\frac{4}{81}} =$ ______ $\sqrt{0.04} =$ ______ $-\sqrt{36} =$ ______ $\sqrt{\frac{4}{9}} =$ ______

F. $\sqrt{0.64} =$ ______ $\sqrt{0.25} =$ ______ $-\sqrt{0.16} =$ ______ $\sqrt{\frac{16}{49}} =$ ______

Use a calculator to find the square root of each number below. Round your answers to the nearest tenth.

G. $\sqrt{56} =$ ______ $\sqrt{13} =$ ______ $\sqrt{91} =$ ______ $\sqrt{21} =$ ______

H. $\sqrt{110} =$ ______ $\sqrt{87} =$ ______ $\sqrt{250} =$ ______ $\sqrt{17} =$ ______

I. $\sqrt{46} =$ ______ $\sqrt{112} =$ ______ $\sqrt{70} =$ ______ $\sqrt{19} =$ ______

J. $\sqrt{57} =$ ______ $\sqrt{83} =$ ______ $\sqrt{7} =$ ______ $\sqrt{96} =$ ______

K. $\sqrt{30} =$ ______ $\sqrt{58} =$ ______ $\sqrt{10} =$ ______ $\sqrt{2} =$ ______

L. $\sqrt{53} =$ ______ $\sqrt{150} =$ ______ $\sqrt{3} =$ ______ $\sqrt{24} =$ ______

Use Two Steps

Solve. Show your work.

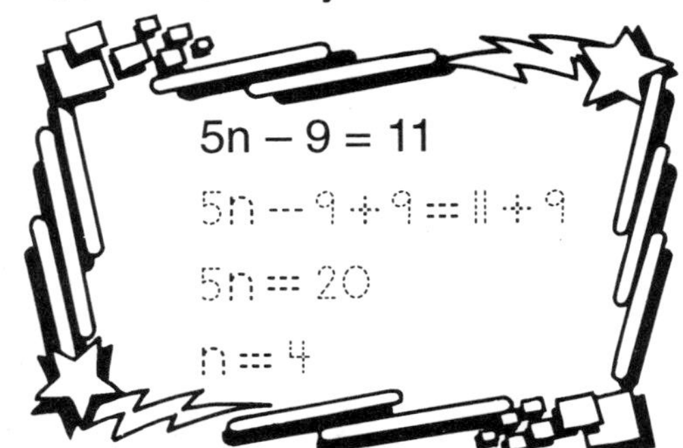

A. $5y + 3 = 18$ $3y - 4 = 14$

B. $2x - 1 = 11$ $3y - 8 = 16$ $4n + 9 = {}^-3$

C. $5c + 7 = {}^-28$ $\frac{x}{2} - 5 = 3$ $\frac{2}{3}b + 4 = 5$

D. $\frac{a}{4} - 2 = 3$ $2x + 5 = 19$ $\frac{x}{2} + 3 = 9$

E. $8y + 11 = 83$ $5n + 2 = {}^-33$ $\frac{1}{5}z - 1 = 3$

Name ____________________

Writing and Solving Complex Equations

Write an equation for each problem. Then solve for the variable.

2 times s decreased by 1 is 9.

$2s - 1 = 9$

$2s = 10$

$s = 5$

A. 4 more than 4 times x is 45.

B. 24 less than triple the value of g is 15.

1 more than e divided by 5 is 2.

C. 3 more than the product of y and 5 is 38.

6 less than the quotient of h divided by 2 is 10.

D. 35 more than twice the value of e is 65.

3 times w increased by 17 is 62.

E. 8 increased by the product of k and 6 is 50.

10 less than the quotient of d divided by 5 is 10.

Name ____________________

Distribute and Solve

Solve the equations.

A. $3(x + 2) = 12$ $\quad$ $4(x - 3) = 8$

B. $2(x + 7) = 20$ $\quad$ $5(x - 5) = 5$ $\quad$ $6(x - 7) = 12$

C. $5(x - 3) = 5$ $\quad$ $7(x - 2) = 28$ $\quad$ $5(x + 2) = 40$

D. $2(x + 8) = 20$ $\quad$ $6(x + 2) = 54$ $\quad$ $3(x - 3) = 30$

E. $4(x + 6) = 60$ $\quad$ $8(x - 4) = 64$ $\quad$ $7(x + 3) = 91$

F. $9(x - 2) = 0$ $\quad$ $5(x + 9) = 100$ $\quad$ $8(x - 8) = 8$

Name ____ Using the distributive property

Distribute and Solve

Solve the equations. Remember to multiply both numbers inside the parentheses by the number in front of the parentheses.

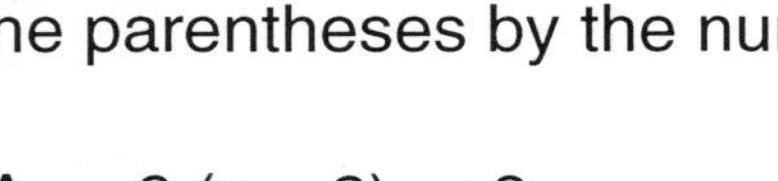

A. $3(x - 2) = {}^{-}3$ $\quad$ $2(x + 5) = {}^{-}2$ $\quad$ ${}^{-}4(x - 5) = 24$

B. ${}^{-}6(x + 7) = 12$ $\quad$ ${}^{-}4(m + 12) = 36$ $\quad$ $2(x - 4) = 22$

C. $20 = 4(x + 3)$ $\quad$ $3(y - 7) = 27$ $\quad$ $3(a - 5) = {}^{-}21$

D. $4(y + 8) = 22$ $\quad$ $35 = 5(n + 2)$ $\quad$ $6(z + 4) = {}^{-}10$

E. ${}^{-}5(a + 7) = 5$ $\quad$ $3(b - 4) = {}^{-}6$ $\quad$ ${}^{-}18 = 3(x - 4)$

F. ${}^{-}6(x - 1) = 12$ $\quad$ ${}^{-}9(x + 1) = 18$ $\quad$ $4(x - 5) = {}^{-}42$

Name ______________________________

Combine the Variables

Multiply the number and variable inside each parentheses by the number in front of the parentheses. Then combine the like variables and solve the equation.

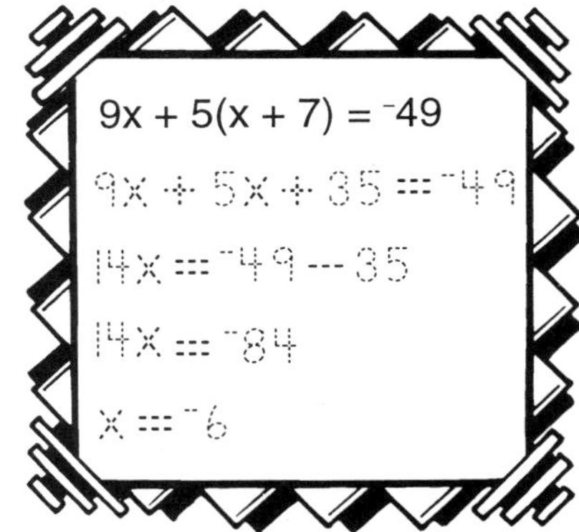

$9x + 5(x + 7) = {}^-49$
$9x + 5x + 35 = {}^-49$
$14x = {}^-49 - 35$
$14x = {}^-84$
$x = {}^-6$

A. $2b + 3(b - 7) = 44$ $\qquad$ $2v + 3(8 - v) = {}^-16$

B. $4(2y + 9) + 7y = {}^-24$ $\qquad$ $5(3n) + 14 + 6n = 21$ $\qquad$ $5(n + 3) + 5 = {}^-25$

C. ${}^-8a + 6(a + 7) = 1$ $\qquad$ $10z + 5(z - 12) = 0$ $\qquad$ $7y + 7(y + 3) = {}^-21$

D. ${}^-3x + 6(x + 4) = 9$ $\qquad$ $4z + 9 + 3(2z) = 129$ $\qquad$ $6(3a - 2) + 5a = 57$

E. $5x + 3(x + 4) = 76$ $\qquad$ ${}^-3(x + 4) - 21x = 72$ $\qquad$ ${}^-4(3 + f) + 6f = {}^-24$

Name ______________________________

On Both Sides Now

Arrange the variables so that they are on the same side of the equation. Then solve the equation.

A. $2x + 72 = 4x$ $\quad$ $24 + y = 9y$

B. $9b = 26 - 4b$ $\quad$ $7n = 15 - 8n$ $\quad$ $12h = 48 - 4h$

C. $42 + 9c = 16c$ $\quad$ $7f = 3f - 52$ $\quad$ $9r + 7 = 4r - 8$

D. $^{-}5w - 9 = 3w + 7$ $\quad$ $72 + 7d = 15d$ $\quad$ $7r = {}^{-}5r + 144$

E. $^{-}3g + 9 = 15g - 9$ $\quad$ $5e = 14 - 2e$ $\quad$ $3s + 10 = 5s + 4$

F. $6(s + 1) = 4s + 8$ $\quad$ $7b + 6 = 2b - 19$ $\quad$ $8x - 1 = 23 - 4x$

Name ____________________

Be an Equation Buster!

Solve each equation. Show your work.

A. $5a + 7 = {}^-23$ $\quad$ $x + 38 = 86 - 3x$

B. $5n = 2n + 6$ $\quad$ $y = 24 - 3y$ $\quad$ ${}^-12n = 35 - 5n$

C. $\frac{{}^-x}{3} = 7$ $\quad$ $4n + 5 = 6n + 7$ $\quad$ $2(x - 6) = 3x$

D. $3p - 8 = 13 - 4p$ $\quad$ $8(5 - n) = 2n$ $\quad$ $5(2 + n) = 3n + 18$

E. $4u - 8 = {}^-5(1 - u)$ $\quad$ $2(x - 6) = 3x$ $\quad$ $\frac{x}{2} + 5 = 6$

F. $98 - 4b = {}^-11b$ $\quad$ $\frac{1}{2}(x + 8) = 10$ $\quad$ $3(m + 5) = 2m + 10$

Name ____________________

More Equation Busting

Solve each equation. Show your work.

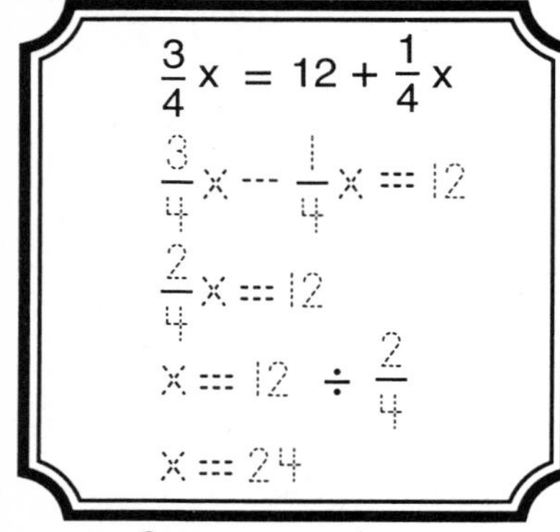

A. $8a = 2a + 30$ $\quad$ $2b = 80 - 8b$

B. $\frac{2}{3}x = 6 - \frac{1}{3}x$ $\quad$ $51 = 9 - 3x$ $\quad$ $39c = 33c - 78$

C. $5p - 9 = 2p + 12$ $\quad$ $89 + x = 2 - 2x$ $\quad$ $4(y - 6) = 7y$

D. $7(10 - m) = 3m$ $\quad$ $\frac{2}{3}x - 7 = x$ $\quad$ $3(30 + s) = 4(s + 19)$

E. $^{-}7a = {}^{-}12a + 65$ $\quad$ $4(x + 2) = 6x + 10$ $\quad$ $\frac{x}{2} + 5 = x$

F. $3(2 + v) = 5v + 16$ $\quad$ $4(3y - 1) = 5y - 11$ $\quad$ $5x + 2(1 - x) = 2(2x - 1)$

Name ____________________

More Combinations

A. How many combinations of two books are possible if you don't want a book about humor or crafts? ________

B. How many combinations of two books are possible if you don't like to read mysteries? ________

C. How many combinations of two books are possible? ________

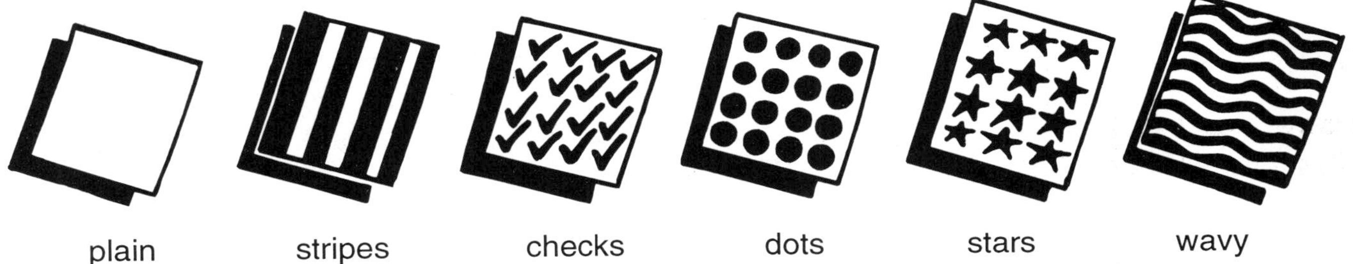

D. How many combinations of two quilt squares are possible? ________

E. How many combinations of two quilt squares are possible if you only use plain, stripes, and stars? ________

F. How many combinations of two quilt squares are possible if you don't use any checks or stripes? ________

G. How many combinations of two quilt squares are possible if you don't use any dots? ________

Name ______________________________

Probability of Simple Events

Look at the spinner in each box. Find the probability of each event. Express your answer as a fraction. **P(N)** means *the probability of getting that number.*

A. P(5) = $\frac{1}{6}$ P(1) = ________

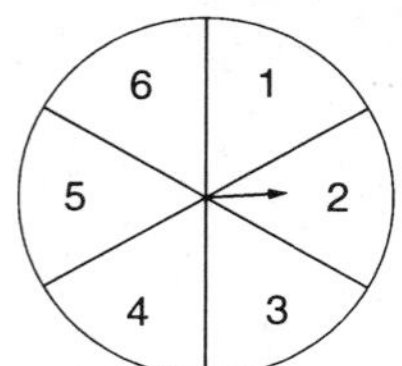

B. P(3 or 4) = ________ P(>5) = ________ P(even number) = ________

C. P(0) = ________ P(<3) = ________ P(<10) = ________

D. P(R) = ________ P(S) = ________

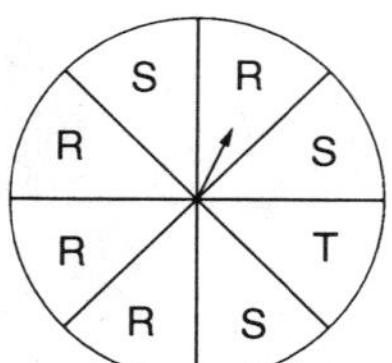

E. P(T) = ________ P(U) = ________

F. P(R or S) = ________ P(S or T) = ________ P(R or T) = ________

G. P(R, S, or T) = ________ P(vowel) = ________ P(not T) = ________

H. P(1) = ________ P(2) = ________

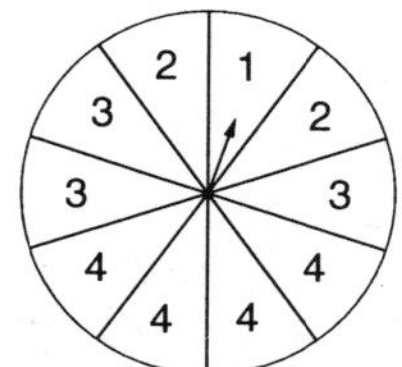

I. P(3) = ________ P(4) = ________

J. P(<3) = ________ P(even number) = ________ P(not 1) = ________

K. P(>1) = ________ P(not 4) = ________ P(1 or 4) = ________

L. P(≥2) = ________ P(2 or 3) = ________ P(not 2) = ________

Name ______________________________

Independent Events

Find the probability of the events in each box. Express your answers as fractions.

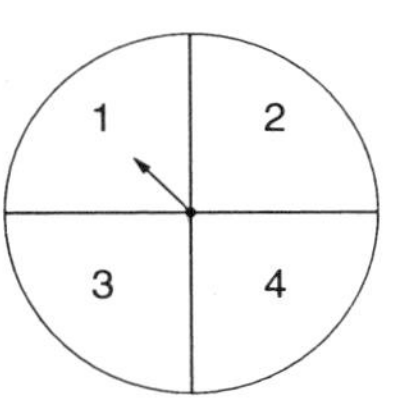

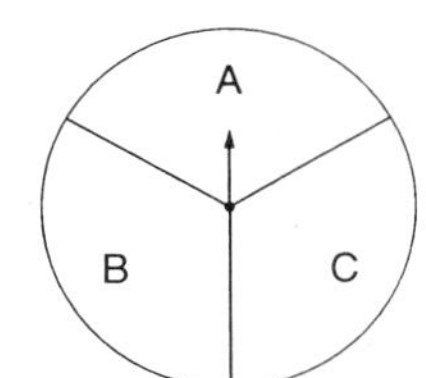

A. P(1, A) = $\frac{1}{4} \bullet \frac{1}{3} = \frac{1}{12}$ P(2, B) = __________

B. P(1, A or B) = __________ P(1 or 2, C) = __________

C. P(1, D) = __________ P(even, A) = __________ P(2, vowel) = __________

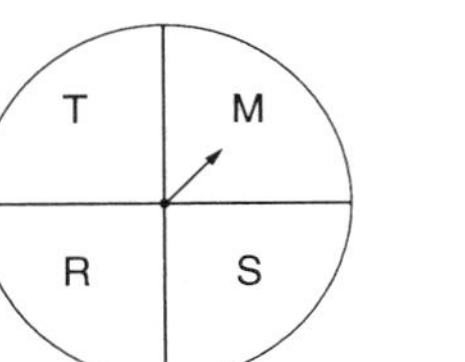

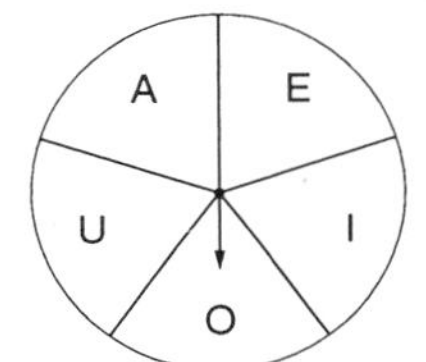

D. P(M, A) = __________ P(T, O) = __________

E. P(M, R) = __________ P(S, vowel) = __________

F. P(T, not A) = __________ P(not R, A) = __________ P(T, A or E) = __________

G. P(T or R, O) = __________ P(R or M, not A) = __________ P(not S, I or O) = __________

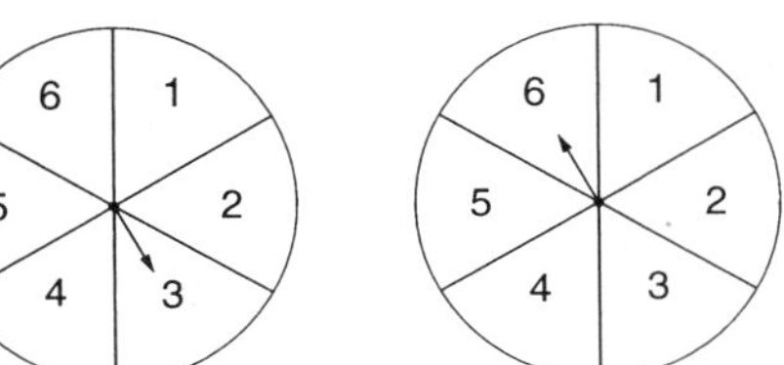

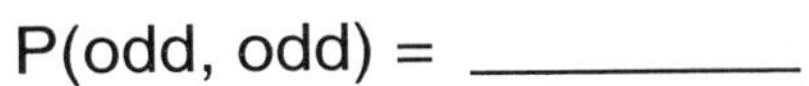

H. P(5, 5) = __________ P(6, 2) = __________

I. P(4, not 2) = __________ P(not 1, 6) = __________

J. P(3, >4) = __________ P(≥4, 1) = __________ P(odd, odd) = __________

K. P(even, odd) = __________ P(even, 3) = __________ P(0, 5) = __________

L. P(not 2, not 3) = __________ P(not 1, even) = __________ P(5, not 3) = __________

Name ____________________ Probability of dependent events

Dependent Events

The bag contains: 2 striped cubes
3 dotted cubes
3 black cubes
4 white cubes

Find the probability of each event if you pick one cube and then pick another without replacing the first.

A. P(white, then dotted) = $\frac{4}{12} \cdot \frac{3}{11} = \frac{4}{44} = \frac{1}{11}$ P(black, then dotted) = ______

B. P(dotted, then black) = ______ P(striped, then white) = ______

C. P(white, then white) = ______ P(dotted, then dotted) = ______

D. P(black, then black) = ______ P(striped, then striped) = ______

E. P(dotted, then white) = ______ P(striped, then black) = ______

Find the probability of each event if you pick one coin and then pick another without replacing the first.

F. P(quarter, then dime) = ______

G. P(dime, then quarter) = ______ P(nickel, then dime) = ______

H. P(nickel, then nickel) = ______ P(penny, then dime) = ______

I. P(dime, then nickel) = ______ P(penny, then penny) = ______

J. P(quarter, then quarter) = ______ P(dime, then dime) = ______

K. P(nickel, then penny) = ______ P(dime, then penny) = ______

Answer Key

Name ______________________ Estimating sums

Make an Estimate

Estimate the sums by rounding the numbers and adding mentally. (Round to the place that makes sense for each number or problem.)

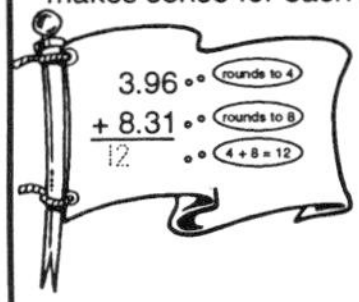

3.96 rounds to 4
+ 8.31 rounds to 8
12 4 + 8 = 12

A.		2.04 + 5.75 = 8	6.014 + 3.275 = 9	10.33 + 17.4 = 27	3.49 + 15.52 = 19
B.	87.43 + 80.40 = 167	1.28 + 9.40 = 10	61.29 + 59.24 = 121	22.14 + 95.52 = 120	2.4 + 39.6 = 42
C.	5.37 + 3.79 = 9	29.34 + 34.56 = 60	70.16 + 45.42 = 115	256.11 + 29.93 = 290	49.99 + 34.56 = 85
D.	24.92 + 76.38 + 67.21 = 170	74.89 + 27.36 + 56.03 = 160	19.55 + 4.46 + 20.26 = 44	124.95 + 59.50 + 100.40 = 280	9.95 + 41.45 + 60.88 = 110
E.	21.07 + 39.54 + 60.24 = 120	70.88 + 12.45 + 3.98 = 84	26.97 + 39.51 + 59.94 = 130	13.06 + 26.91 + 5.27 = 45	434.15 + 202.01 + 46.82 = 700

* Estimates will vary. Possible estimates are shown.

Page 1

Name ______________________ Estimating differences

Rough Estimates

Estimate the differences by rounding the numbers and subtracting mentally. (Round to the place that makes sense for each number or problem.)

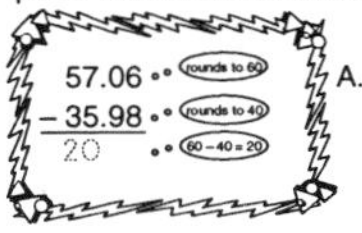

57.06 rounds to 60
− 35.98 rounds to 40
20 60 − 40 = 20

A.		9.67 − 3.81 = 6	5.03 − 4.21 = 1	11.20 − 9.67 = 1	7.83 − 1.96 = 6
B.	12.40 − 10.66 = 1	19.05 − 4.41 = 15	32.79 − 12.19 = 21	12.89 − 5.24 = 8	9.36 − 5.44 = 4
C.	69.05 − 19.55 = 50	10.71 − 10.69 = 0	52.74 − 13.14 = 40	46.83 − 20.91 = 25	84.37 − 79.66 = 4
D.	55.70 − 51.42 = 5	68.30 − 45.75 = 20	8.13 − 7.95 = 0	32.78 − 18.29 = 10	7.83 − 6.10 = 2
E.	42.90 − 19.72 = 20	18.05 − 8.68 = 9	9.46 − 1.78 = 7	32.13 − 1.97 = 28	42.73 − 10.99 = 32
F.	237.94 − 99.75 = 138	481.3 − 76.1 = 400	47.85 − 19.99 = 28	9.65 − 5.84 = 4	10.039 − 9.876 = 0

* Estimates will vary. Possible estimates are shown.

Page 2

Name ______________________ Adding decimals

Sum It Up

Find the sums.

A.	6.2 + 1.73 = 7.93	0.525 + 0.139 = 0.644	12.36 + 8.75 = 21.11	7.65 + 3.34 = 10.99	0.68 + 0.4 = 1.08
B.	72.88 + 14.75 = 87.63	9.89 + 42.69 = 52.58	0.25 + 0.73 = 0.98	5.9 + 6.057 = 11.957	38.93 + 17.6 = 56.53
C.	4.769 + 3.825 = 8.594	78.9 + 32.6 = 111.5	34.9 + 67.85 = 102.75	1.79 + 3.826 = 5.616	36.29 + 28.75 = 65.04
D.	4.7 + 8.8 + 0.45 = 13.95	439.6 + 7.049 + 12.32 = 458.969	72.6 + 123.12 + 2.4 = 198.12	2.36 + 3.78 + 2.67 = 8.81	42.83 + 75.6 + 36.356 = 154.786
E.	7.84 + 65.3 + 238.72 = 311.86	179.6 + 4.98 + 56.43 = 241.01	2.368 + 3.26 + 0.471 = 6.099	27.6 + 3.84 + 51.09 = 82.53	

Page 3

Name ______________________ Subtracting decimals

Decimal Differences

Find the differences.

A.	265.3 − 121.44 = 143.86	3.74 − 1.88 = 1.86	52.67 − 24.7 = 27.97	57.19 − 19.88 = 37.31	
B.	5.25 − 3.87 = 1.38	0.85 − 0.68 = 0.17	51.04 − 22.63 = 28.41	70.00 − 16.95 = 53.05	143.79 − 88.81 = 54.98
C.	26.85 − 15.97 = 10.88	71.35 − 4.661 = 66.689	73.21 − 56.56 = 16.65	54.135 − 27.950 = 26.185	95.41 − 8.72 = 86.69
D.	23.9 − 18.72 = 5.18	44.04 − 28.15 = 15.89	1.343 − 0.975 = 0.368	680.3 − 136.9 = 543.4	7.342 − 2.617 = 4.725
E.	56.53 − 17.6 = 38.93	213.06 − 4.8 = 208.26	1.04 − 0.999 = 0.041	8.35 − 2.967 = 5.383	4.0 − 2.91 = 1.09
F.	439.3 − 97.42 = 341.88	83.81 − 7.96 = 75.85	100.1 − 83.79 = 16.31	17.31 − 14.9 = 2.41	6.01 − 3.07 = 2.94

Page 4

Answer Key

Name ____________________ Estimating products

Product Estimates

Estimate the products by rounding the numbers and multiplying. (Round to the place that makes sense for each number or problem.)

8.19 (rounds to 8)
x 3.9 (rounds to 4)
32 (8 x 4 = 32)

A.	32.9 x 8.6 270	43.82 x 17.9 880	8.97 x 56.3 560	27.87 x 3.56 120	
B.	52.7 x 1.87 106	19.97 x 4.95 100	3.7 x 8.3 32	6.47 x 7.19 42	8.96 x 4.1 36
C.	71.974 x 8.6 720	67.46 x 39.28 2,800	319.3 x 86.76 27,000	909.36 x 6.2 5,400	3.826 x 7.4 28
D.	9.15 x 57.8 580	6.45 x 8.59 54	7.83 x 0.84 8	48.42 x 2.26 96	8.99 x 2.3 18
E.	3.62 x 7.8 32	26.9 x 7.3 300	43.46 x 37.3 1,600	808.2 x 5.8 4,800	29.5 x 0.137 3
F.	63.40 x 4.91 300	29.2 x 8.71 270	73.9 x 2.03 148	57.8 x 6.34 360	98.4 x 0.5 98

* Estimates will vary. Possible estimates are shown.

Page 5

Name ____________________ Multiplying decimals

Multiplying Decimals

Find the products.

A.	3.5 x 6.7 23.45	8.09 x 5.7 46.113	12.5 x 0.74 9.25	9.4 x 2.7 25.38	12.8 x 3.5 44.8
B.	5.12 x 7.6 38.912	9.12 x 6.8 62.016	0.73 x 4.2 3.066	5.6 x 8.3 46.48	6.9 x 5.4 37.26
C.	8.42 x 7.3 61.466	7.58 x 4.8 36.384	53.7 x 6.9 370.53	4.86 x 3.7 17.982	6.45 x 7.6 49.02
D.	7.25 x 1.89 13.7025	5.62 x 3.84 21.5808	3.79 x 1.01 3.8279	1.23 x 3.7 4.551	

Use estimation to see if your answers are reasonable.

Page 6

Name ____________________ Estimating quotients

Quotient Estimates

Estimate the quotients by rounding the numbers and dividing. (Round to the place that makes sense for each number or problem.)

176.2 ÷ 6.3
That's close to 180 ÷ 6, so the quotient will be about 30.

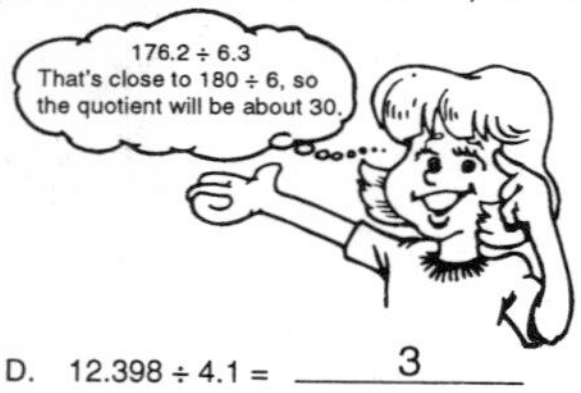

A. 176.2 ÷ 63 = 3

B. 301.38 ÷ 5.3 = 60

C. 11.93 ÷ 3.2 = 4

D.	12.398 ÷ 4.1 = 3		162.58 ÷ 4.4 = 40
E.	482.04 ÷ 8.32 = 60		14.948 ÷ 5.1 = 3
F.	359.6 ÷ 8.5 = 40		266.7 ÷ 3.1 = 90
G.	64.26 ÷ 8.3 = 8		12.345 ÷ 3.3 = 4
H.	354.64 ÷ 69.3 = 5		1,615.9 ÷ 4.35 = 400
I.	413.66 ÷ 5.1 = 80		279.31 ÷ 4.1 = 70
J.	3,112.4 ÷ 5.2 = 600		363.63 ÷ 3.6 = 91
K.	409.4 ÷ 48.4 = 8		123.94 ÷ 3.9 = 31
L.	54.43 ÷ 9.32 = 6		34.594 ÷ 4.8 = 7
M.	48.14 ÷ 6.4 = 8		561.84 ÷ 79.3 = 7
N.	544.82 ÷ 9.1 = 60		355.39 ÷ 5.7 = 60

* Estimates will vary. Possible estimates are shown.

Page 7

Name ____________________ Dividing with decimals

Decimal Division

Find the quotients.

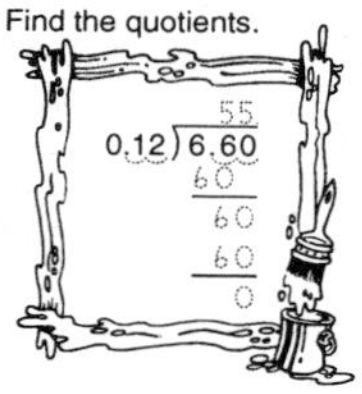

55
0.12)6.60
60
60
60
0

A.	65 0.4)26	58 0.5)29	72 0.46)33.12	
B.	40 0.25)10	3.8 3.4)12.92	6.3 0.67)4.221	9.8 0.03)0.294
C.	0.087 9.5)0.8265	32 0.7)22.4	18.2 0.08)1.456	5 2.1)10.5
D.	46 0.08)3.68	26 0.12)3.12	3.5 0.42)1.470	6.5 8.4)54.6

Page 8

Answer Key

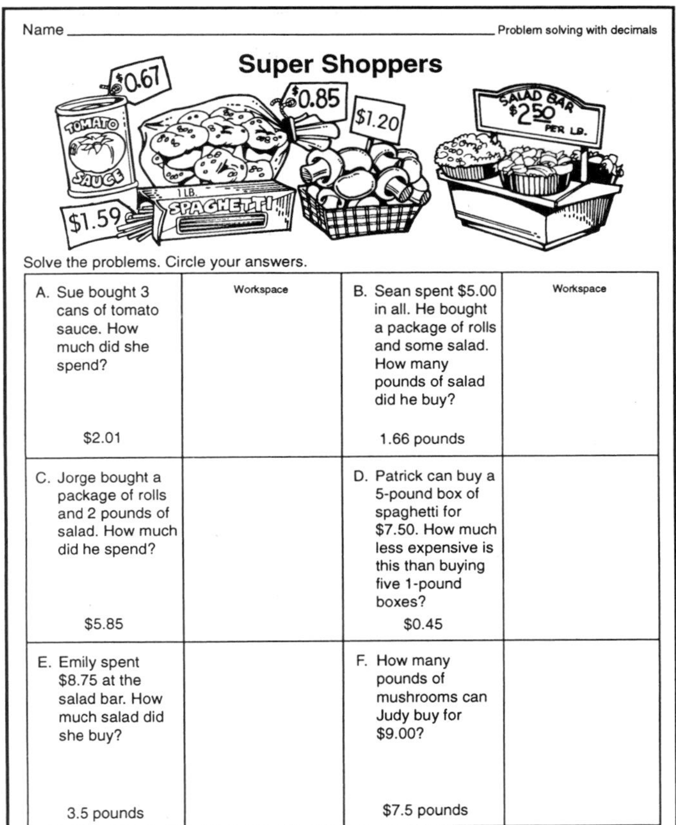

Name ________________ Problem solving with decimals

Super Shoppers

Solve the problems. Circle your answers.

	Workspace		Workspace
A. Sue bought 3 cans of tomato sauce. How much did she spend? $2.01		B. Sean spent $5.00 in all. He bought a package of rolls and some salad. How many pounds of salad did he buy? 1.66 pounds	
C. Jorge bought a package of rolls and 2 pounds of salad. How much did he spend? $5.85		D. Patrick can buy a 5-pound box of spaghetti for $7.50. How much less expensive is this than buying five 1-pound boxes? $0.45	
E. Emily spent $8.75 at the salad bar. How much salad did she buy? 3.5 pounds		F. How many pounds of mushrooms can Judy buy for $9.00? $7.5 pounds	

Page 9

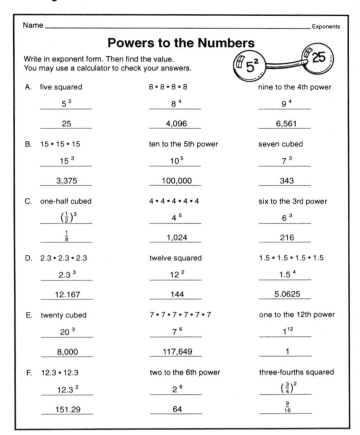

Name ________________ Exponents

Powers to the Numbers

Write in exponent form. Then find the value.
You may use a calculator to check your answers.

5^2 — 25

A.	five squared: 5^2 = 25	8 • 8 • 8 • 8: 8^4 = 4,096	nine to the 4th power: 9^4 = 6,561
B.	15 • 15 • 15: 15^3 = 3,375	ten to the 5th power: 10^5 = 100,000	seven cubed: 7^3 = 343
C.	one-half cubed: $(\frac{1}{2})^3$ = $\frac{1}{8}$	4 • 4 • 4 • 4 • 4: 4^5 = 1,024	six to the 3rd power: 6^3 = 216
D.	2.3 • 2.3 • 2.3: 2.3^3 = 12.167	twelve squared: 12^2 = 144	1.5 • 1.5 • 1.5 • 1.5: 1.5^4 = 5.0625
E.	twenty cubed: 20^3 = 8,000	7 • 7 • 7 • 7 • 7 • 7: 7^6 = 117,649	one to the 12th power: 1^{12} = 1
F.	12.3 • 12.3: 12.3^2 = 151.29	two to the 6th power: 2^6 = 64	three-fourths squared: $(\frac{3}{4})^2$ = $\frac{9}{16}$

Page 10

Name ________________ Scientific notation

Scientific Notation

Scientific notation is an easy way to represent a large number. To write a number in scientific notation, move the decimal point the number of places necessary to get a number from 1 to 9. Then write the number from 1 to 9 with a multiplication sign and a power of 10 showing the number of places you moved the decimal point.

> $8{,}700{,}000 = 8.7 \times 10^6$
> 8.7 is between 1 and 9.
> You moved the decimal point 6 places.

A.	5,300: 5.3×10^3	70,000: 7×10^4	260: 2.6×10^2	890,000: 8.9×10^5
B.	450: 4.5×10^2	8,200: 8.2×10^3	45,300: 4.53×10^4	976,000: 9.76×10^5
C.	416,000: 4.16×10^5	52,000: 5.2×10^4	93,700: 9.37×10^4	4,960: 4.96×10^3

Write the numbers in standard form.

D.	3.5×10^2: 350	4.37×10^4: 43,700	9.7×10^5: 970,000	1.315×10^3: 1,315
E.	3.4×10^3: 3,400	8.37×10^5: 837,000	1.05×10^2: 105	9.876×10^4: 98,760
F.	2.46×10^6: 2,460,000	9.21×10^4: 92,100	3.89×10^6: 3,890,000	8.4×10^8: 840,000,000

Page 11

Name ________________ Divisibility

Even Steven

A number is divisible by

2	3	4	5	6	9	10
if it ends in 0, 2, 4, 6, or 8	if the sum of the digits is divisible by 3	if the last two digits form a number that is divisible by 4	if it ends in 0 or 5	if it is divisible by 2 and 3	if the sum of the digits is divisible by 9	if it ends in 0

Complete the table. Write **Y** (yes) or **N** (no).

	Divisible by						
Number	2	3	4	5	6	9	10
540	y	y	y	y	y	y	y
346	y	n	n	n	n	n	n
621	n	y	n	n	n	y	n
2,690	y	n	n	y	n	n	y
5,211	n	y	n	n	n	y	n
4,002	y	y	n	n	y	n	n
6,732	y	y	y	n	y	y	n
9,017	n	n	n	n	n	n	n
10,950	y	y	n	y	y	n	y
12,579	n	y	n	n	n	n	n
34,782	y	y	n	n	y	n	n
56,712	y	y	y	n	y	n	n
67,125	n	y	n	y	n	n	n
79,470	y	y	n	y	y	y	y

Page 12

Answer Key

Name ______ Prime and composite numbers

Primo Primes

A prime number has only two whole-number factors—itself and 1.
A composite number has more than two whole-number factors.
Write **prime** or **composite** beside each number.

A.	3	prime	41	prime		
B.	18	composite	19	prime		
C.	64	composite	39	composite	52	composite
D.	10	composite	11	prime	15	composite
E.	6	composite	17	prime	25	composite
F.	5	prime	51	composite	36	composite
G.	49	composite	7	prime	91	composite
H.	53	prime	85	composite	13	prime
I.	89	prime	23	prime	93	composite
J.	43	prime	79	prime	47	prime
K.	73	prime	75	composite	86	composite
L.	67	prime	83	prime	29	prime
M.	31	prime	87	composite	97	prime
N.	94	composite	59	prime	37	prime
O.	61	prime	81	composite	63	composite

Page 13

Name ______ Factoring numbers

Perfect Products

List all of the factors of each number from least to greatest.
Then tell whether the number of factors is odd or even.

	Number	Factors	Odd or Even Number of Factors
A.	12	1, 2, 3, 4, 6, 12	Even
B.	16	1, 2, 4, 8, 16	Odd
C.	18	1, 2, 3, 6, 9, 18	Even
D.	20	1, 2, 4, 5, 10, 20	Even
E.	25	1, 5, 25	Odd
F.	32	1, 2, 4, 8, 16, 32	Even
G.	36	1, 2, 3, 4, 6, 9, 12, 18, 36	Odd
H.	40	1, 2, 4, 5, 8, 10, 20, 40	Even
I.	48	1, 2, 3, 4, 6, 8, 12, 16, 24, 48	Even
J.	56	1, 2, 4, 7, 8, 14, 28, 56	Even
K.	60	1, 2, 3, 4, 5, 6, 10, 12, 15, 20, 30, 60	Even
L.	64	1, 2, 4, 8, 16, 32, 64	Odd
M.	72	1, 2, 3, 4, 6, 8, 9, 12, 18, 24, 36, 72	Even
N.	81	1, 3, 9, 27, 81	Odd
O.	100	1, 2, 4, 5, 10, 20, 25, 50, 100	Odd
P.	121	1, 11, 121	Odd
Q.	144	1, 2, 3, 4, 6, 8, 9, 12, 16, 18, 24, 36, 48, 72, 144	Odd
R.	225	1, 3, 5, 9, 15, 25, 45, 75, 225	Odd

Look at the numbers that have an odd number of factors. The middle factor of each number should be the square root of the number.

Page 14

Name ______ Prime factorization with exponents

Find the Prime Factors

Draw a factor tree to find the prime factors. Then write the prime factors using exponents.

A. 75: $5^2 \cdot 3$ 88: $2^3 \cdot 11$ 54: $3^3 \cdot 2$

B. 20: $5 \cdot 2^2$ 50: $5^2 \cdot 2$ 36: $2^2 \cdot 3^2$

C. 98: $7^2 \cdot 2$ 90: $3^2 \cdot 5 \cdot 2$ 120: $3 \cdot 2^3 \cdot 5$

D. 60: $5 \cdot 3 \cdot 2^2$ 32: 2^5

Page 15

Name ______ Greatest common factor

Factors Are Great!

On scratch paper, find the factors for each number.
Write the **greatest common factor** (GCF) for each pair of numbers on the line below the pair.

Factors of 18: 9, 18 | Both: 1, 2, 3, 6 | Factors of 24: 4, 8, 12, 24
GCF = 6

A.	16 and 40: 8	10 and 21: 1	
B.	24 and 40: 8	36 and 54: 9	48 and 64: 16
C.	18 and 27: 9	12 and 36: 12	21 and 28: 7
D.	16 and 24: 8	18 and 30: 6	8 and 27: 1
E.	45 and 60: 15	28 and 42: 14	48 and 72: 24
F.	26 and 51: 1	100 and 130: 10	24 and 72: 24
G.	27 and 81: 27	18 and 32: 2	42 and 56: 14
H.	90 and 189: 9	91 and 95: 1	84 and 108: 12
I.	144 and 216: 72	136 and 162: 2	121 and 143: 11

Page 16

Answer Key

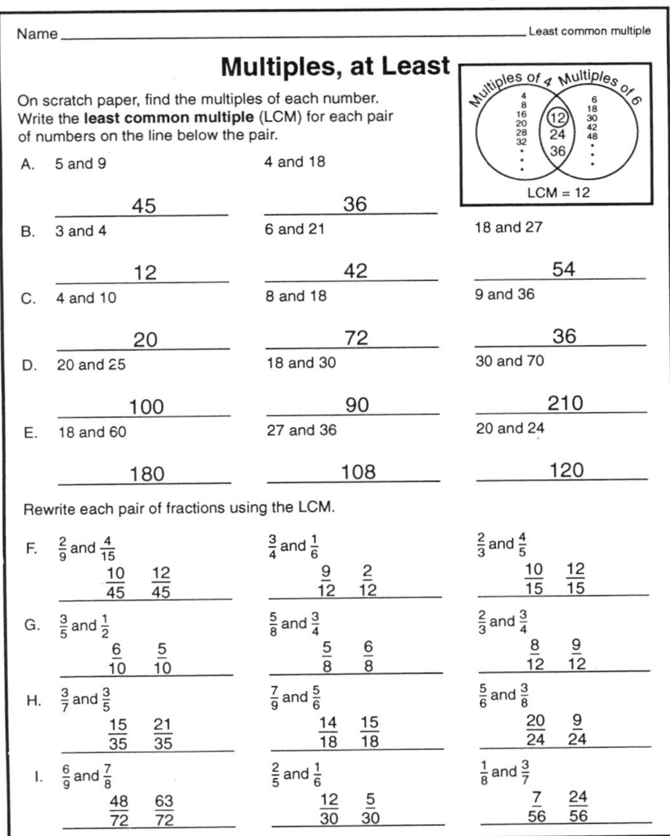

Name ____________ Least common multiple

Multiples, at Least

On scratch paper, find the multiples of each number. Write the **least common multiple** (LCM) for each pair of numbers on the line below the pair.

A. 5 and 9: 45; 4 and 18: 36

B. 3 and 4: 12; 6 and 21: 42; 18 and 27: 54

C. 4 and 10: 20; 8 and 18: 72; 9 and 36: 36

D. 20 and 25: 100; 18 and 30: 90; 30 and 70: 210

E. 18 and 60: 180; 27 and 36: 108; 20 and 24: 120

Rewrite each pair of fractions using the LCM.

F. $\frac{2}{9}$ and $\frac{4}{15}$: $\frac{10}{45}$ $\frac{12}{45}$; $\frac{3}{4}$ and $\frac{1}{6}$: $\frac{9}{12}$ $\frac{2}{12}$; $\frac{2}{3}$ and $\frac{4}{5}$: $\frac{10}{15}$ $\frac{12}{15}$

G. $\frac{3}{5}$ and $\frac{1}{2}$: $\frac{6}{10}$ $\frac{5}{10}$; $\frac{5}{8}$ and $\frac{3}{4}$: $\frac{5}{8}$ $\frac{6}{8}$; $\frac{2}{3}$ and $\frac{3}{4}$: $\frac{8}{12}$ $\frac{9}{12}$

H. $\frac{3}{7}$ and $\frac{3}{5}$: $\frac{15}{35}$ $\frac{21}{35}$; $\frac{7}{9}$ and $\frac{5}{6}$: $\frac{14}{18}$ $\frac{15}{18}$; $\frac{5}{6}$ and $\frac{3}{8}$: $\frac{20}{24}$ $\frac{9}{24}$

I. $\frac{6}{9}$ and $\frac{7}{8}$: $\frac{48}{72}$ $\frac{63}{72}$; $\frac{2}{5}$ and $\frac{1}{6}$: $\frac{12}{30}$ $\frac{5}{30}$; $\frac{1}{8}$ and $\frac{3}{7}$: $\frac{7}{56}$ $\frac{24}{56}$

Page 17

Name ____________ Comparing and ordering fractions

Order These

Write in order from least to greatest. Use scratch paper if necessary.

Sometimes I use a number line and sometimes I use the LCM.

A. $\frac{1}{5}, \frac{1}{3}, \frac{1}{4}$ → $\frac{1}{5}, \frac{1}{4}, \frac{1}{3}$; $\frac{4}{9}, \frac{1}{2}, \frac{2}{3}$ → $\frac{4}{9}, \frac{1}{2}, \frac{2}{3}$; $\frac{2}{5}, \frac{7}{15}, \frac{1}{2}$ → $\frac{2}{5}, \frac{7}{15}, \frac{1}{2}$

B. $\frac{5}{6}, \frac{6}{7}, \frac{3}{21}$ → $\frac{3}{8}, \frac{5}{6}, \frac{6}{7}$; $\frac{5}{8}, \frac{1}{2}, \frac{3}{4}$ → $\frac{1}{2}, \frac{5}{8}, \frac{3}{4}$; $\frac{4}{5}, \frac{1}{2}, \frac{9}{10}$ → $\frac{1}{2}, \frac{4}{5}, \frac{9}{10}$

C. $\frac{2}{5}, \frac{1}{2}, \frac{3}{10}$ → $\frac{3}{10}, \frac{2}{5}, \frac{1}{2}$; $\frac{3}{5}, \frac{1}{2}, \frac{1}{4}$ → $\frac{1}{4}, \frac{1}{2}, \frac{3}{5}$; $\frac{3}{8}, \frac{2}{3}, \frac{1}{4}$ → $\frac{1}{4}, \frac{3}{8}, \frac{2}{3}$; $\frac{7}{15}, \frac{2}{3}, \frac{1}{5}$ → $\frac{1}{5}, \frac{7}{15}, \frac{2}{3}$

D. $\frac{3}{4}, \frac{3}{5}, \frac{7}{10}$ → $\frac{3}{5}, \frac{7}{10}, \frac{3}{4}$; $\frac{1}{3}, \frac{3}{8}, \frac{1}{4}$ → $\frac{1}{4}, \frac{1}{3}, \frac{3}{8}$; $\frac{2}{5}, \frac{4}{9}, \frac{11}{15}$ → $\frac{2}{5}, \frac{4}{9}, \frac{11}{15}$; $\frac{5}{6}, \frac{3}{8}, \frac{1}{2}$ → $\frac{3}{8}, \frac{1}{2}, \frac{5}{6}$

E. $\frac{2}{5}, \frac{1}{3}, \frac{1}{4}$ → $\frac{1}{4}, \frac{1}{3}, \frac{2}{5}$; $\frac{2}{9}, \frac{5}{18}, \frac{3}{6}$ → $\frac{2}{9}, \frac{5}{18}, \frac{3}{6}$; $\frac{5}{6}, \frac{4}{5}, \frac{3}{4}$ → $\frac{3}{4}, \frac{4}{5}, \frac{5}{6}$; $\frac{1}{9}, \frac{1}{7}, \frac{1}{8}$ → $\frac{1}{9}, \frac{1}{8}, \frac{1}{7}$

F. $\frac{2}{3}, \frac{5}{6}, \frac{3}{4}$ → $\frac{2}{3}, \frac{3}{4}, \frac{5}{6}$; $\frac{1}{2}, \frac{3}{10}, \frac{4}{9}$ → $\frac{3}{10}, \frac{4}{9}, \frac{1}{2}$; $\frac{5}{6}, \frac{2}{5}, \frac{2}{3}$ → $\frac{2}{5}, \frac{2}{3}, \frac{5}{6}$; $\frac{2}{7}, \frac{2}{5}, \frac{3}{10}$ → $\frac{2}{7}, \frac{3}{10}, \frac{2}{5}$

G. $\frac{5}{6}, \frac{7}{12}, \frac{2}{3}$ → $\frac{7}{12}, \frac{2}{3}, \frac{5}{6}$; $\frac{1}{5}, \frac{1}{9}, \frac{1}{3}$ → $\frac{1}{9}, \frac{1}{5}, \frac{1}{3}$; $\frac{2}{7}, \frac{2}{11}, \frac{2}{5}$ → $\frac{2}{11}, \frac{2}{7}, \frac{2}{5}$; $\frac{1}{3}, \frac{3}{10}, \frac{2}{5}$ → $\frac{3}{10}, \frac{1}{3}, \frac{2}{5}$

Page 18

Name ____________ Adding and subtracting fractions

Watch the Signs

Add or subtract. Write your answer in the simplest form.

A. $\frac{2}{3} + \frac{1}{3} = \frac{3}{3} = 1$; $\frac{1}{8} + \frac{4}{8} = \frac{5}{8}$; $\frac{6}{11} + \frac{2}{11} = \frac{8}{11}$; $\frac{10}{13} - \frac{4}{13} = \frac{6}{13}$

B. $\frac{11}{12} - \frac{5}{12} = \frac{6}{12} = \frac{1}{2}$; $\frac{4}{9} + \frac{2}{9} = \frac{6}{9} = \frac{2}{3}$; $\frac{2}{15} + \frac{8}{15} = \frac{10}{15} = \frac{2}{3}$; $\frac{3}{20} + \frac{11}{20} = \frac{14}{20} = \frac{7}{10}$

C. $\frac{2}{5} + \frac{1}{2} = \frac{9}{10}$; $\frac{7}{12} + \frac{1}{4} = \frac{10}{12} = \frac{5}{6}$; $\frac{9}{10} - \frac{1}{2} = \frac{4}{10} = \frac{2}{5}$; $\frac{2}{3} - \frac{1}{6} = \frac{3}{6} = \frac{1}{2}$

D. $\frac{13}{18} - \frac{2}{9} = \frac{9}{18} = \frac{1}{2}$; $\frac{7}{24} + \frac{5}{12} = \frac{17}{24}$; $\frac{5}{8} - \frac{4}{7} = \frac{3}{56}$; $\frac{14}{15} + \frac{8}{9} = \frac{82}{45} = 1\frac{37}{45}$

E. $\frac{4}{9} + \frac{3}{4} = \frac{43}{36} = 1\frac{7}{36}$; $\frac{2}{5} + \frac{4}{15} = \frac{10}{15} = \frac{2}{3}$; $\frac{1}{3} - \frac{1}{7} = \frac{4}{21}$; $\frac{4}{5} - \frac{1}{3} = \frac{7}{15}$

F. $\frac{1}{9} - \frac{1}{12} = \frac{1}{36}$; $\frac{3}{10} + \frac{5}{6} + \frac{2}{5} = \frac{46}{30} = 1\frac{8}{15}$; $\frac{4}{9} + \frac{1}{3} + \frac{5}{6} = \frac{29}{18} = 1\frac{11}{18}$

Page 19

Name ____________ Adding mixed numbers

Mixed Number Sums

Add. Write your answer in the simplest form.

I have to find the least common denominator first!

A. $3\frac{1}{2} + 2\frac{1}{6} = 5\frac{2}{3}$; $7\frac{3}{10} + 9\frac{3}{4} = 17\frac{1}{20}$

B. $12\frac{5}{6} + 6\frac{7}{9} = 19\frac{11}{18}$; $6\frac{4}{7} + 2\frac{9}{14} = 9\frac{3}{14}$; $8\frac{7}{12} + 6\frac{5}{8} = 15\frac{5}{24}$; $11\frac{1}{6} + 7\frac{1}{2} = 18\frac{2}{3}$

C. $1\frac{5}{6} + 6\frac{1}{2} = 8\frac{1}{3}$; $7\frac{2}{3} + 6\frac{3}{5} = 14\frac{4}{15}$; $4\frac{1}{4} + 7\frac{7}{8} = 12\frac{1}{8}$; $3\frac{2}{3} + 2\frac{2}{3} = 6\frac{1}{3}$

D. $6\frac{1}{6} + 1\frac{11}{12} = 8\frac{1}{12}$; $3\frac{3}{4} + 6\frac{1}{2} = 10\frac{1}{4}$; $2\frac{5}{6} + 5\frac{3}{4} = 8\frac{7}{12}$; $2\frac{1}{8} + 8\frac{2}{3} = 10\frac{19}{24}$

E. $3\frac{1}{3} + 4\frac{5}{6} + 1\frac{1}{12} = 9\frac{1}{4}$; $6\frac{2}{3} + 1\frac{1}{3} + 3\frac{1}{2} = 11\frac{1}{2}$; $5\frac{1}{2} + 2\frac{1}{3} + 11\frac{1}{6} = 19$; $2\frac{1}{2} + 1\frac{5}{8} + 3\frac{3}{4} = 7\frac{7}{8}$

F. $5\frac{1}{4} + 9\frac{1}{12} + 4\frac{1}{6} = 18\frac{1}{2}$; $2\frac{1}{15} + 11\frac{2}{5} + 3\frac{1}{3} = 16\frac{4}{5}$; $7\frac{1}{2} + 14\frac{1}{10} + 4\frac{2}{5} = 26$; $3\frac{2}{3} + 1\frac{1}{2} + 4\frac{3}{4} = 9\frac{11}{12}$

Page 20

Answer Key

Name ____________________ Subtracting mixed numbers

Mixed Number Differences

Subtract. Write your answer in the simplest form.

A.	$8\frac{4}{5} - 4\frac{1}{2} = 4\frac{3}{10}$	$7\frac{3}{10} - 6\frac{3}{4} = \frac{11}{20}$	$9\frac{7}{9} - 5\frac{1}{3} = 4\frac{4}{9}$	$8\frac{3}{4} - 2\frac{5}{8} = 6\frac{1}{8}$
B.	$6\frac{5}{6} - 3\frac{7}{9} = 3\frac{1}{18}$	$2\frac{1}{9} - \frac{1}{2} = 1\frac{11}{18}$	$18\frac{1}{3} - 9 = 9\frac{1}{3}$	$17\frac{3}{5} - 6\frac{1}{3} = 11\frac{4}{15}$
C.	$4\frac{9}{10} - 3\frac{2}{5} = 1\frac{1}{2}$	$4\frac{5}{8} - 1\frac{1}{4} = 3\frac{3}{8}$	$4\frac{7}{9} - 2\frac{4}{9} = 2\frac{1}{3}$	$3\frac{3}{10} - 2 = 1\frac{3}{10}$
D.	$5\frac{9}{10} - 4\frac{3}{5} = 1\frac{3}{10}$	$3 - 1\frac{7}{10} = 1\frac{3}{10}$	$6\frac{3}{8} - 2\frac{3}{16} = 4\frac{3}{16}$	$4\frac{1}{10} - 2\frac{3}{10} = 1\frac{4}{5}$
E.	$8\frac{1}{4} - 4\frac{3}{4} = 3\frac{1}{2}$	$12\frac{2}{5} - 8\frac{9}{10} = 3\frac{1}{2}$	$15\frac{1}{2} - 9\frac{9}{16} = 5\frac{15}{16}$	$12\frac{9}{10} - \frac{17}{20} = 12\frac{1}{20}$
F.	$4\frac{1}{2} - 2\frac{3}{4} = 1\frac{3}{4}$	$14\frac{1}{9} - 9\frac{2}{3} = 4\frac{4}{9}$	$9\frac{1}{3} - 6\frac{1}{2} = 2\frac{5}{6}$	

Did you remember to find the least common denominator? Did you regroup when you needed to?

Page 21

Name ____________________ Adding and subtracting fractions

Fraction Practice

Add or subtract. Write your answers in the simplest form.

A.	$\frac{1}{5} + \frac{3}{10} = \frac{1}{2}$	$\frac{3}{4} - \frac{3}{8} = \frac{3}{8}$	$4\frac{7}{8} + 4\frac{7}{8} = 9\frac{3}{4}$	
B.	$\frac{4}{5} + 1\frac{5}{6} = 1\frac{49}{30} = 2\frac{19}{30}$	$9\frac{1}{8} + 6\frac{3}{4} = 15\frac{7}{8}$	$16\frac{5}{11} - 9 = 7\frac{5}{11}$	$6\frac{4}{7} + 2\frac{1}{5} = 8\frac{27}{35}$
C.	$13\frac{1}{3} + 9\frac{8}{17} = 22\frac{41}{51}$	$7 - 6\frac{1}{8} = \frac{7}{8}$	$8\frac{1}{12} + 3\frac{1}{3} = 11\frac{5}{12}$	$5\frac{9}{14} - 3\frac{6}{7} = 1\frac{11}{14}$
D.	$6 - 1\frac{19}{21} = 4\frac{2}{21}$	$6\frac{7}{11} + 3 = 9\frac{7}{11}$	$4\frac{1}{2} + 10\frac{11}{14} = 15\frac{2}{7}$	$9\frac{3}{7} - 4\frac{1}{6} = 5\frac{11}{42}$
E.	$\frac{5}{12} + 4\frac{5}{6} = 5\frac{1}{4}$	$15\frac{1}{7} - 6\frac{2}{3} = 8\frac{10}{21}$	$5\frac{9}{10} + 1\frac{1}{2} = 7\frac{2}{5}$	$9\frac{1}{4} - 3\frac{2}{5} = 5\frac{17}{20}$
F.	$\frac{11}{14} + 1\frac{1}{7} + 3\frac{1}{4} = 5\frac{5}{28}$	$6\frac{1}{6} + 5\frac{1}{3} + 10\frac{1}{2} = 22$	$19\frac{1}{2} - 9\frac{3}{4} = 9\frac{3}{4}$	$2\frac{3}{8} + 5\frac{1}{6} + 3\frac{11}{12} = 11\frac{11}{24}$

Page 22

Name ____________________ Multiplying fractions

Products From Fractions

Multiply. Divide any numerator and denominator by a common factor to make the fractions easier to multiply. Write your answers in the simplest form.

$\frac{1}{3} \cdot \frac{3}{5} = \frac{1}{5}$

A.		$\frac{1}{4} \cdot \frac{4}{9} = \frac{1}{9}$	$\frac{7}{8} \cdot \frac{20}{21} = \frac{5}{6}$
B.		$\frac{7}{8} \cdot \frac{8}{7} = 1$	$\frac{9}{16} \cdot \frac{10}{9} = \frac{5}{8}$
C.	$\frac{7}{11} \cdot \frac{23}{42} = \frac{23}{66}$	$\frac{3}{8} \cdot \frac{4}{5} = \frac{3}{10}$	$\frac{5}{3} \cdot \frac{2}{5} = \frac{2}{3}$
D.	$\frac{3}{2} \cdot \frac{5}{6} = \frac{5}{4} = 1\frac{1}{4}$	$\frac{3}{4} \cdot \frac{2}{9} = \frac{1}{6}$	$\frac{2}{3} \cdot \frac{6}{5} = \frac{4}{5}$
E.	$\frac{5}{7} \cdot \frac{7}{10} = \frac{1}{2}$	$\frac{4}{5} \cdot \frac{1}{8} = \frac{1}{10}$	$\frac{7}{12} \cdot \frac{3}{7} = \frac{1}{4}$
F.	$\frac{4}{7} \cdot \frac{7}{2} = 2$	$\frac{3}{5} \cdot \frac{25}{6} = 2\frac{1}{2}$	$\frac{4}{5} \cdot \frac{3}{2} = 1\frac{1}{5}$
G.	$\frac{7}{8} \cdot \frac{4}{21} = \frac{1}{6}$	$\frac{2}{11} \cdot \frac{11}{24} = \frac{1}{12}$	$\frac{1}{3} \cdot \frac{3}{8} = \frac{1}{8}$
H.	$\frac{3}{5} \cdot \frac{2}{3} = \frac{2}{5}$	$\frac{6}{7} \cdot \frac{5}{12} = \frac{5}{14}$	$\frac{4}{5} \cdot \frac{2}{7} = \frac{8}{35}$
I.	$\frac{7}{10} \cdot \frac{5}{8} = \frac{7}{16}$	$\frac{2}{7} \cdot \frac{7}{8} = \frac{1}{4}$	$\frac{6}{7} \cdot \frac{14}{15} = \frac{4}{5}$
J.	$\frac{3}{4} \cdot \frac{8}{9} = \frac{2}{3}$	$\frac{8}{5} \cdot \frac{15}{16} = 1\frac{1}{2}$	$\frac{9}{10} \cdot \frac{4}{3} = 1\frac{1}{5}$
K.	$\frac{5}{2} \cdot \frac{1}{2} = 1\frac{1}{4}$	$\frac{5}{3} \cdot \frac{15}{8} = \frac{25}{8} = 3\frac{1}{8}$	$\frac{4}{3} \cdot \frac{5}{8} = \frac{5}{6}$
L.	$\frac{1}{3} \cdot \frac{3}{5} \cdot \frac{5}{7} = \frac{1}{7}$	$\frac{5}{2} \cdot \frac{2}{3} \cdot \frac{3}{10} = \frac{1}{2}$	$\frac{3}{4} \cdot \frac{4}{5} \cdot \frac{1}{3} = \frac{1}{5}$

Page 23

Name ____________________ Multiplying mixed numbers

Mixed Number Multiplication

Find the products. Rewrite mixed numbers as improper fractions before you multiply. Write your answers in the simplest form.

A.	$16 \cdot \frac{3}{4} = \frac{12}{1} = 12$	$20 \cdot \frac{3}{5} = 12$	
B.	$6\frac{2}{3} \cdot \frac{1}{4} = \frac{5}{3} = 1\frac{2}{3}$	$5\frac{1}{8} \cdot 6 = \frac{123}{4} = 30\frac{3}{4}$	$3 \cdot 5\frac{1}{2} = \frac{33}{2} = 16\frac{1}{2}$
C.	$7\frac{1}{7} \cdot \frac{3}{8} = \frac{75}{28} = 2\frac{19}{28}$	$7\frac{3}{4} \cdot 20 = 155$	$4\frac{3}{8} \cdot \frac{2}{5} = \frac{7}{4} = 1\frac{3}{4}$
D.	$1\frac{1}{3} \cdot 30 = 40$	$2\frac{1}{2} \cdot \frac{5}{6} = \frac{25}{12} = 2\frac{1}{12}$	$9\frac{2}{9} \cdot \frac{18}{25} = \frac{166}{25} = 6\frac{16}{25}$
E.	$\frac{3}{8} \cdot 7\frac{2}{3} = \frac{23}{8} = 2\frac{7}{8}$	$5\frac{1}{4} \cdot 16 = 84$	$7\frac{1}{3} \cdot \frac{6}{11} = 4$
F.	$2\frac{2}{3} \cdot 4\frac{1}{2} = 12$	$9\frac{7}{8} \cdot 2\frac{2}{3} = 26\frac{1}{3}$	$12\frac{1}{2} \cdot \frac{4}{5} = 10$
G.	$4\frac{2}{11} \cdot 22 = 92$	$18 \cdot 7\frac{4}{9} = 134$	$\frac{1}{2} \cdot 24 = 12$
H.	$4\frac{2}{5} \cdot 25 = 110$	$5\frac{1}{3} \cdot 9\frac{1}{8} = \frac{146}{3} = 48\frac{2}{3}$	$15 \cdot 8\frac{1}{3} = 125$
I.	$15 \cdot 9\frac{2}{3} = 145$	$2\frac{1}{4} \cdot 11\frac{1}{3} = \frac{102}{4} = 25\frac{1}{2}$	$5\frac{1}{3} \cdot 1\frac{1}{4} = \frac{20}{3} = 6\frac{2}{3}$
J.	$2\frac{3}{4} \cdot \frac{1}{8} = \frac{11}{32}$	$3\frac{1}{3} \cdot 4\frac{2}{5} = \frac{44}{3} = 14\frac{2}{3}$	$2\frac{5}{8} \cdot 16 = 42$
K.	$3\frac{3}{4} \cdot 16 = 60$	$\frac{1}{9} \cdot 3\frac{2}{3} = \frac{11}{27}$	$6\frac{2}{3} \cdot 24 = 160$

Page 24

Answer Key

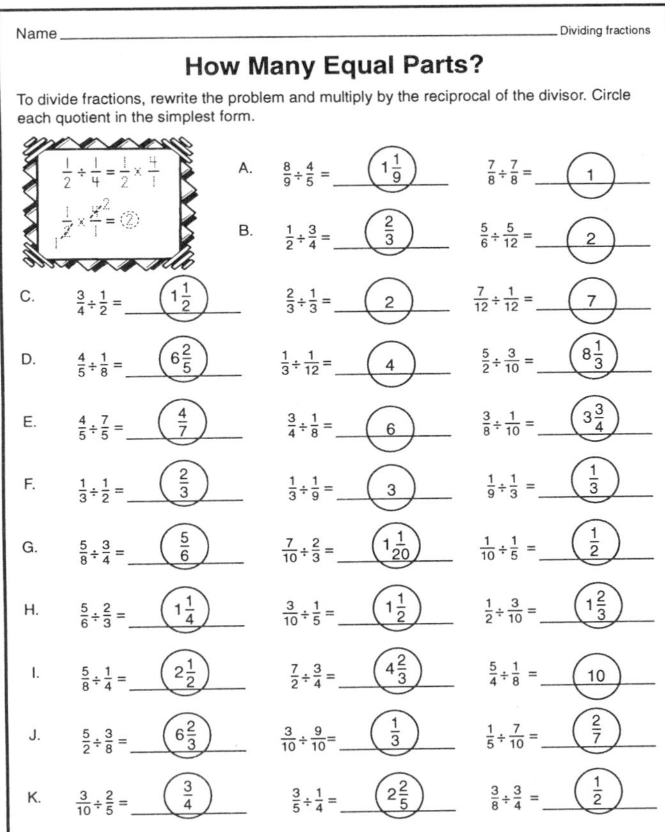

Name ______________________ Dividing fractions

How Many Equal Parts?

To divide fractions, rewrite the problem and multiply by the reciprocal of the divisor. Circle each quotient in the simplest form.

$\frac{1}{2} \div \frac{1}{4} = \frac{1}{2} \times \frac{4}{1}$

$\frac{1}{2} \times \frac{4}{1} = 2$

A. $\frac{8}{9} \div \frac{4}{5} = 1\frac{1}{9}$ $\frac{7}{8} \div \frac{7}{8} = 1$

B. $\frac{1}{2} \div \frac{3}{4} = \frac{2}{3}$ $\frac{5}{6} \div \frac{5}{12} = 2$

C. $\frac{3}{4} \div \frac{1}{2} = 1\frac{1}{2}$ $\frac{2}{3} \div \frac{1}{3} = 2$ $\frac{7}{12} \div \frac{1}{12} = 7$

D. $\frac{4}{5} \div \frac{1}{8} = 6\frac{2}{5}$ $\frac{1}{3} \div \frac{1}{12} = 4$ $\frac{5}{2} \div \frac{3}{10} = 8\frac{1}{3}$

E. $\frac{4}{5} \div \frac{7}{5} = \frac{4}{7}$ $\frac{3}{4} \div \frac{1}{8} = 6$ $\frac{3}{8} \div \frac{1}{10} = 3\frac{3}{4}$

F. $\frac{1}{3} \div \frac{1}{2} = \frac{2}{3}$ $\frac{1}{3} \div \frac{1}{9} = 3$ $\frac{1}{9} \div \frac{1}{3} = \frac{1}{3}$

G. $\frac{5}{8} \div \frac{3}{4} = \frac{5}{6}$ $\frac{7}{10} \div \frac{2}{3} = 1\frac{1}{20}$ $\frac{1}{10} \div \frac{1}{5} = \frac{1}{2}$

H. $\frac{5}{6} \div \frac{2}{3} = 1\frac{1}{4}$ $\frac{3}{10} \div \frac{1}{5} = 1\frac{1}{2}$ $\frac{1}{2} \div \frac{3}{10} = 1\frac{2}{3}$

I. $\frac{5}{8} \div \frac{1}{4} = 2\frac{1}{2}$ $\frac{7}{2} \div \frac{3}{4} = 4\frac{2}{3}$ $\frac{5}{4} \div \frac{1}{8} = 10$

J. $\frac{5}{2} \div \frac{3}{8} = 6\frac{2}{3}$ $\frac{3}{10} \div \frac{9}{10} = \frac{1}{3}$ $\frac{1}{5} \div \frac{7}{10} = \frac{2}{7}$

K. $\frac{3}{10} \div \frac{2}{5} = \frac{3}{4}$ $\frac{3}{5} \div \frac{1}{4} = 2\frac{2}{5}$ $\frac{3}{8} \div \frac{3}{4} = \frac{1}{2}$

Page 25

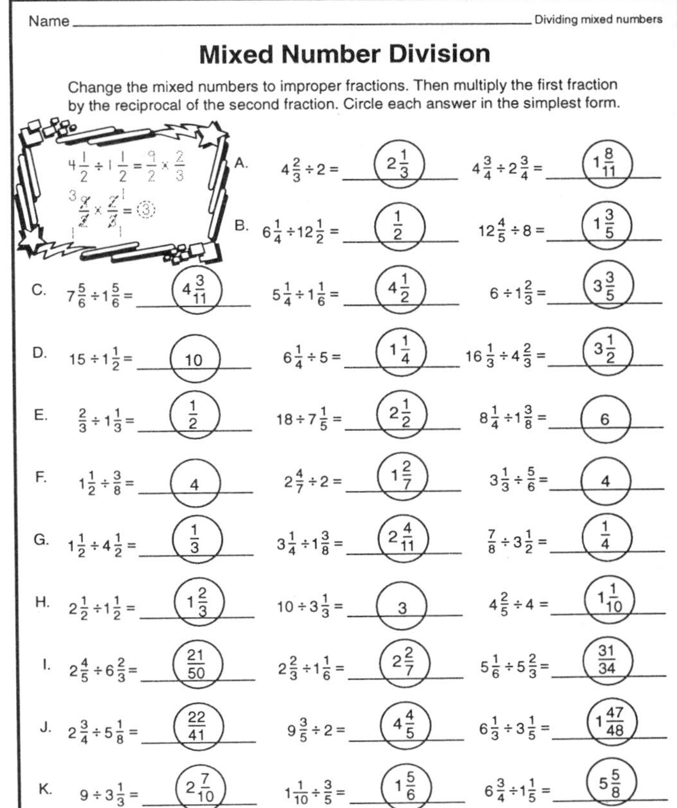

Name ______________________ Dividing mixed numbers

Mixed Number Division

Change the mixed numbers to improper fractions. Then multiply the first fraction by the reciprocal of the second fraction. Circle each answer in the simplest form.

$4\frac{1}{2} \div 1\frac{1}{2} = \frac{9}{2} \times \frac{2}{3}$

$\frac{9}{2} \times \frac{2}{3} = 3$

A. $4\frac{2}{3} \div 2 = 2\frac{1}{3}$ $4\frac{3}{4} \div 2\frac{3}{4} = 1\frac{8}{11}$

B. $6\frac{1}{4} \div 12\frac{1}{2} = \frac{1}{2}$ $12\frac{4}{5} \div 8 = 1\frac{3}{5}$

C. $7\frac{5}{6} \div 1\frac{5}{6} = 4\frac{3}{11}$ $5\frac{1}{4} \div 1\frac{1}{6} = 4\frac{1}{2}$ $6 \div 1\frac{2}{3} = 3\frac{3}{5}$

D. $15 \div 1\frac{1}{2} = 10$ $6\frac{1}{4} \div 5 = 1\frac{1}{4}$ $16\frac{1}{3} \div 4\frac{2}{3} = 3\frac{1}{2}$

E. $\frac{2}{3} \div 1\frac{1}{3} = \frac{1}{2}$ $18 \div 7\frac{1}{5} = 2\frac{1}{2}$ $8\frac{1}{4} \div 1\frac{3}{8} = 6$

F. $1\frac{1}{2} \div \frac{3}{8} = 4$ $2\frac{4}{7} \div 2 = 1\frac{2}{7}$ $3\frac{1}{3} \div \frac{5}{6} = 4$

G. $1\frac{1}{2} \div 4\frac{1}{2} = \frac{1}{3}$ $3\frac{1}{4} \div 1\frac{3}{8} = 2\frac{4}{11}$ $\frac{7}{8} \div 3\frac{1}{2} = \frac{1}{4}$

H. $2\frac{1}{2} \div 1\frac{1}{2} = 1\frac{2}{3}$ $10 \div 3\frac{1}{3} = 3$ $4\frac{2}{5} \div 4 = 1\frac{1}{10}$

I. $2\frac{4}{5} \div 6\frac{2}{3} = \frac{21}{50}$ $2\frac{2}{3} \div 1\frac{1}{6} = 2\frac{2}{7}$ $5\frac{1}{6} \div 5\frac{2}{3} = \frac{31}{34}$

J. $2\frac{3}{4} \div 5\frac{1}{8} = \frac{22}{41}$ $9\frac{3}{5} \div 2 = 4\frac{4}{5}$ $6\frac{1}{3} \div 3\frac{1}{5} = 1\frac{47}{48}$

K. $9 \div 3\frac{1}{3} = 2\frac{7}{10}$ $1\frac{1}{10} \div \frac{3}{5} = 1\frac{5}{6}$ $6\frac{3}{4} \div 1\frac{1}{5} = 5\frac{5}{8}$

Page 26

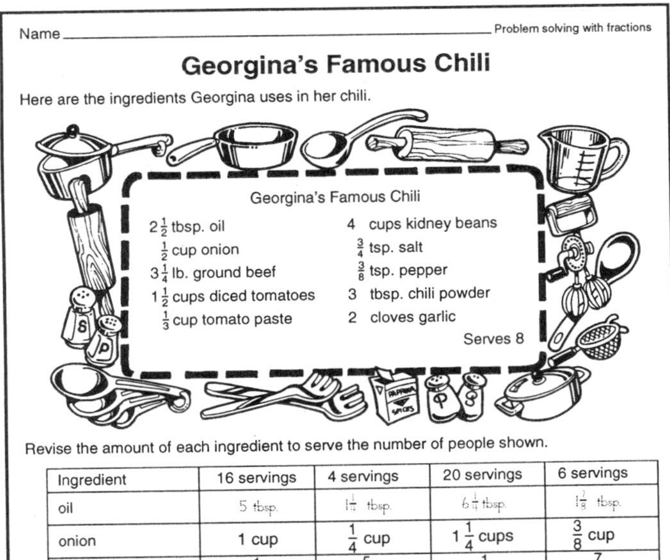

Name ______________________ Problem solving with fractions

Georgina's Famous Chili

Here are the ingredients Georgina uses in her chili.

Georgina's Famous Chili

$2\frac{1}{2}$ tbsp. oil
$\frac{1}{2}$ cup onion
$3\frac{1}{4}$ lb. ground beef
$1\frac{1}{2}$ cups diced tomatoes
$\frac{1}{3}$ cup tomato paste
4 cups kidney beans
$\frac{3}{4}$ tsp. salt
$\frac{3}{8}$ tsp. pepper
3 tbsp. chili powder
2 cloves garlic

Serves 8

Revise the amount of each ingredient to serve the number of people shown.

Ingredient	16 servings	4 servings	20 servings	6 servings
oil	5 tbsp.	$1\frac{1}{4}$ tbsp.	$6\frac{1}{4}$ tbsp.	$1\frac{7}{8}$ tbsp.
onion	1 cup	$\frac{1}{4}$ cup	$1\frac{1}{4}$ cups	$\frac{3}{8}$ cup
ground beef	$6\frac{1}{2}$ lb.	$1\frac{5}{8}$ lb.	$8\frac{1}{8}$ lb.	$2\frac{7}{16}$ lb.
tomatoes	3 cups	$\frac{3}{4}$ cup	$3\frac{3}{4}$ cups	$1\frac{1}{8}$ cups
tomato paste	$\frac{2}{3}$ cup	$\frac{1}{6}$ cup	$\frac{5}{6}$ cup	$\frac{1}{4}$ cup
kidney beans	8 cups	2 cups	10 cups	3 cups
salt	$1\frac{1}{2}$ tsp.	$\frac{3}{8}$ tsp.	$1\frac{7}{8}$ tsp.	$\frac{9}{16}$ tsp.
pepper	$\frac{3}{4}$ tsp.	$\frac{3}{16}$ tsp.	$\frac{15}{16}$ tsp.	$\frac{9}{32}$ tsp.
chili powder	6 tbsp.	$1\frac{1}{2}$ tbsp.	$7\frac{1}{2}$ tbsp.	$2\frac{1}{4}$ tbsp.
garlic	4 cloves	1 cloves	5 cloves	$1\frac{1}{2}$ cloves

Page 27

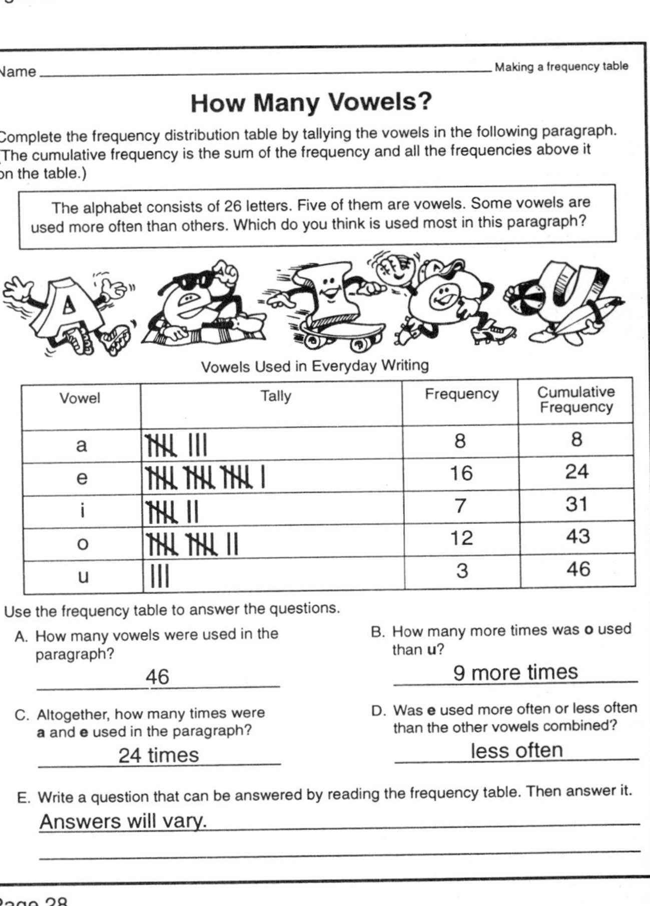

Name ______________________ Making a frequency table

How Many Vowels?

Complete the frequency distribution table by tallying the vowels in the following paragraph. (The cumulative frequency is the sum of the frequency and all the frequencies above it on the table.)

The alphabet consists of 26 letters. Five of them are vowels. Some vowels are used more often than others. Which do you think is used most in this paragraph?

Vowels Used in Everyday Writing

Vowel	Tally	Frequency	Cumulative Frequency
a	𝍸 III	8	8
e	𝍸 𝍸 𝍸 I	16	24
i	𝍸 II	7	31
o	𝍸 𝍸 II	12	43
u	III	3	46

Use the frequency table to answer the questions.

A. How many vowels were used in the paragraph? 46

B. How many more times was **o** used than **u**? 9 more times

C. Altogether, how many times were **a** and **e** used in the paragraph? 24 times

D. Was **e** used more often or less often than the other vowels combined? less often

E. Write a question that can be answered by reading the frequency table. Then answer it.
Answers will vary.

Page 28

Answer Key

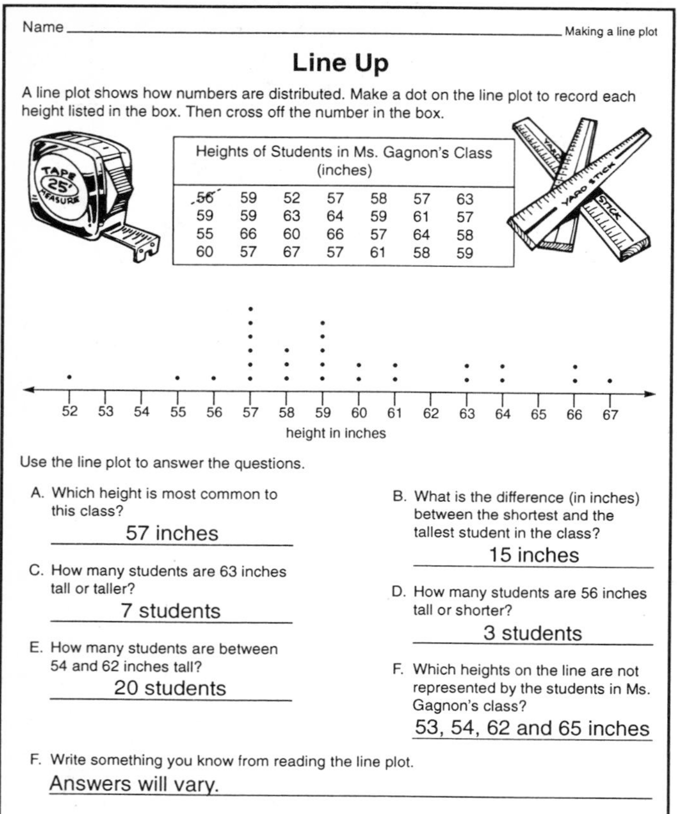

Name ______________________ Making a line plot

Line Up

A line plot shows how numbers are distributed. Make a dot on the line plot to record each height listed in the box. Then cross off the number in the box.

Heights of Students in Ms. Gagnon's Class (inches)						
56	59	52	57	58	57	63
59	59	63	64	59	61	57
55	66	60	66	57	64	58
60	57	67	57	61	58	59

Use the line plot to answer the questions.

A. Which height is most common to this class?
57 inches

B. What is the difference (in inches) between the shortest and the tallest student in the class?
15 inches

C. How many students are 63 inches tall or taller?
7 students

D. How many students are 56 inches tall or shorter?
3 students

E. How many students are between 54 and 62 inches tall?
20 students

F. Which heights on the line are not represented by the students in Ms. Gagnon's class?
53, 54, 62 and 65 inches

F. Write something you know from reading the line plot.
Answers will vary.

Page 29

Name ______________________ Making a stem-and-leaf plot

Stem-and-Leaf Plots

Make a stem-and-leaf plot to organize the test scores listed in the box. In the *stem* column, write the tens digits of the scores in order from least to greatest. In the *leaves* column, write the digits that go with the tens digits in order from least to greatest. Cross off each number as you record it.

Test Scores for Ms. Woo's Math Class			
56	59	83	68
63	67	95	78
74	95	74	92
82	87	88	95
65	88	86	89
76	79	80	76
82	90	90	92

Stem	Leaves
5	6, 9
6	3, 5, 7, 8
7	4, 4, 6, 6, 8, 9
8	0, 2, 2, 3, 6, 7, 8, 8, 9
9	0, 0, 2, 2, 5, 5, 5

Use your stem-and-leaf plot to answer the questions.

A. What was the lowest score?
56

B. What was the highest score?
95

C. How many students scored 90 or above?
7 students

D. How many students scored below 70?
6 students

E. How many students scored between 70 and 89, including 89?
15 students

F. What single score was the most common in Ms. Woo's class?
95

G. Use information from the stem-and-leaf plot and the grading scale at the right to find the number of students who received each letter grade. Write the numbers on the grade record at the right.

Grade Record

Score	Grade	Number of Students
90–100	A	7
80–89	B	9
70–79	C	6
60–69	D	4
1–59	F	2

Page 30

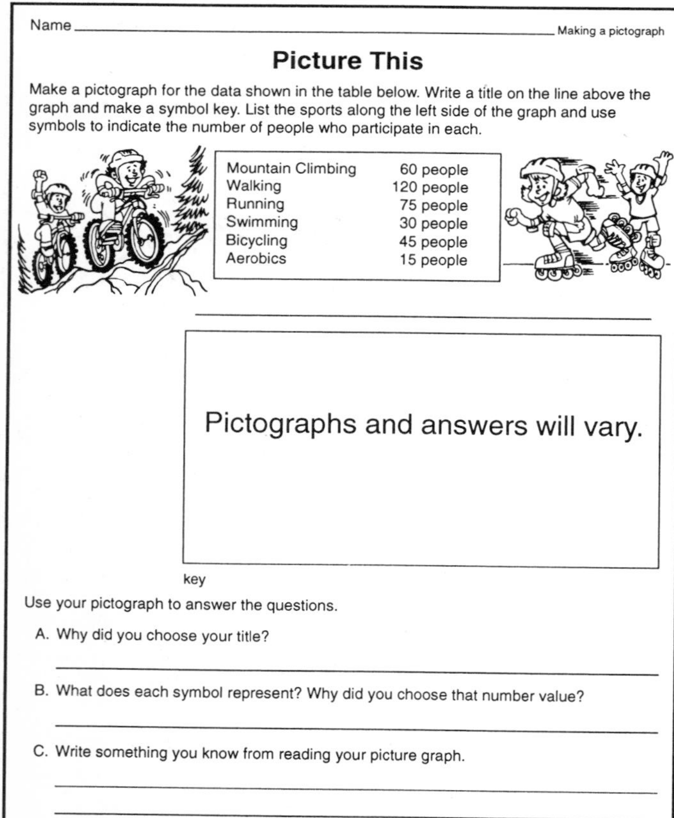

Name ______________________ Making a pictograph

Picture This

Make a pictograph for the data shown in the table below. Write a title on the line above the graph and make a symbol key. List the sports along the left side of the graph and use symbols to indicate the number of people who participate in each.

Mountain Climbing	60 people
Walking	120 people
Running	75 people
Swimming	30 people
Bicycling	45 people
Aerobics	15 people

Pictographs and answers will vary.

key

Use your pictograph to answer the questions.

A. Why did you choose your title?

B. What does each symbol represent? Why did you choose that number value?

C. Write something you know from reading your picture graph.

Page 31

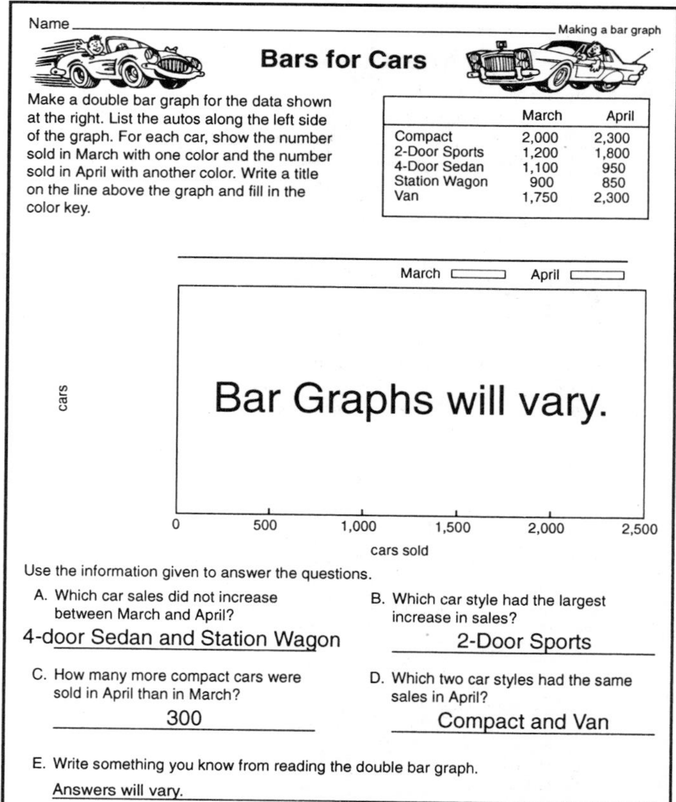

Name ______________________ Making a bar graph

Bars for Cars

Make a double bar graph for the data shown at the right. List the autos along the left side of the graph. For each car, show the number sold in March with one color and the number sold in April with another color. Write a title on the line above the graph and fill in the color key.

	March	April
Compact	2,000	2,300
2-Door Sports	1,200	1,800
4-Door Sedan	1,100	950
Station Wagon	900	850
Van	1,750	2,300

Use the information given to answer the questions.

A. Which car sales did not increase between March and April?
4-door Sedan and Station Wagon

B. Which car style had the largest increase in sales?
2-Door Sports

C. How many more compact cars were sold in April than in March?
300

D. Which two car styles had the same sales in April?
Compact and Van

E. Write something you know from reading the double bar graph.
Answers will vary.

Page 32

Answer Key

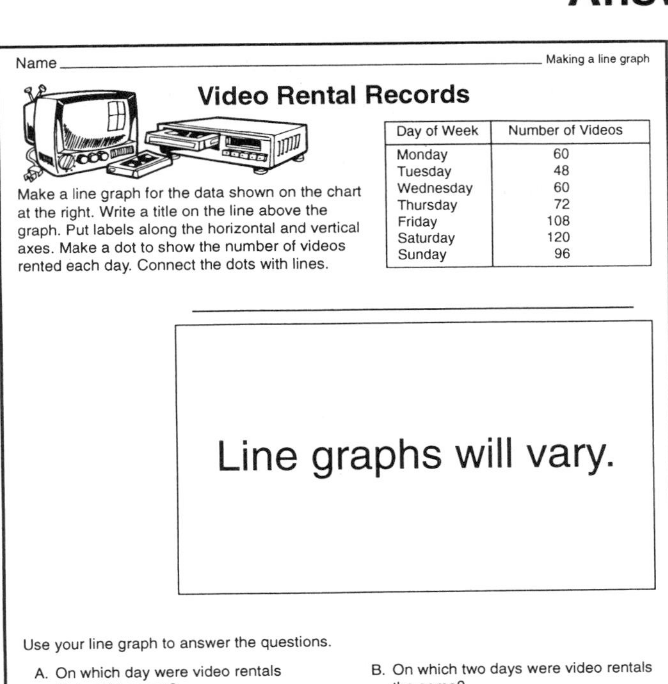

Name ______________________ Making a line graph

Video Rental Records

Day of Week	Number of Videos
Monday	60
Tuesday	48
Wednesday	60
Thursday	72
Friday	108
Saturday	120
Sunday	96

Make a line graph for the data shown on the chart at the right. Write a title on the line above the graph. Put labels along the horizontal and vertical axes. Make a dot to show the number of videos rented each day. Connect the dots with lines.

Line graphs will vary.

Use your line graph to answer the questions.

A. On which day were video rentals highest? Lowest?
Saturday, Tuesday

B. On which two days were video rentals the same?
Monday and Wednesday

C. Did video rentals increase or decrease between Tuesday and Wednesday?
increase

D. What does the graph tell about the trend of video rentals during the course of the week?
Rentals increase after Tuesday, peak on Saturday, and then decrease again, bottoming out on Tuesday

Page 33

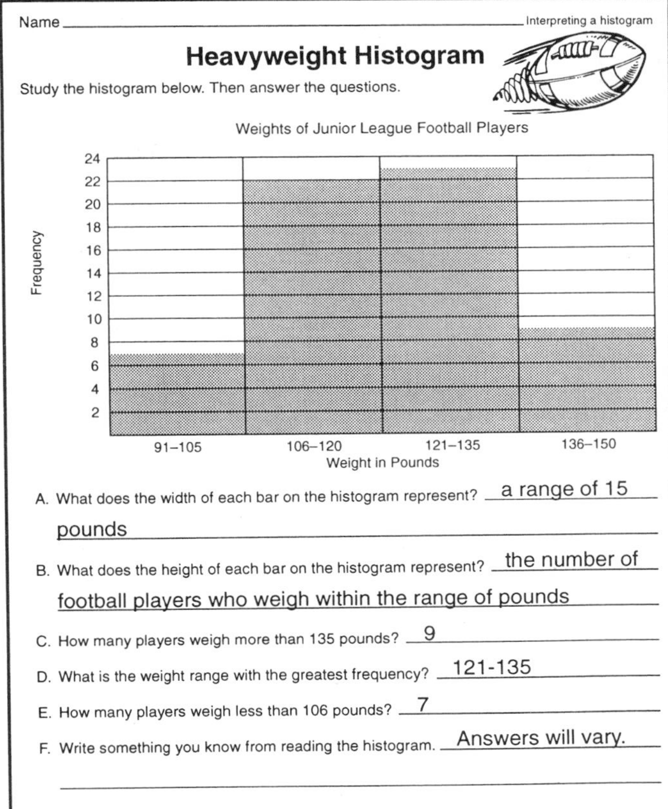

Name ______________________ Interpreting a histogram

Heavyweight Histogram

Study the histogram below. Then answer the questions.

A. What does the width of each bar on the histogram represent? a range of 15 pounds

B. What does the height of each bar on the histogram represent? the number of football players who weigh within the range of pounds

C. How many players weigh more than 135 pounds? 9

D. What is the weight range with the greatest frequency? 121-135

E. How many players weigh less than 106 pounds? 7

F. Write something you know from reading the histogram. Answers will vary.

Page 34

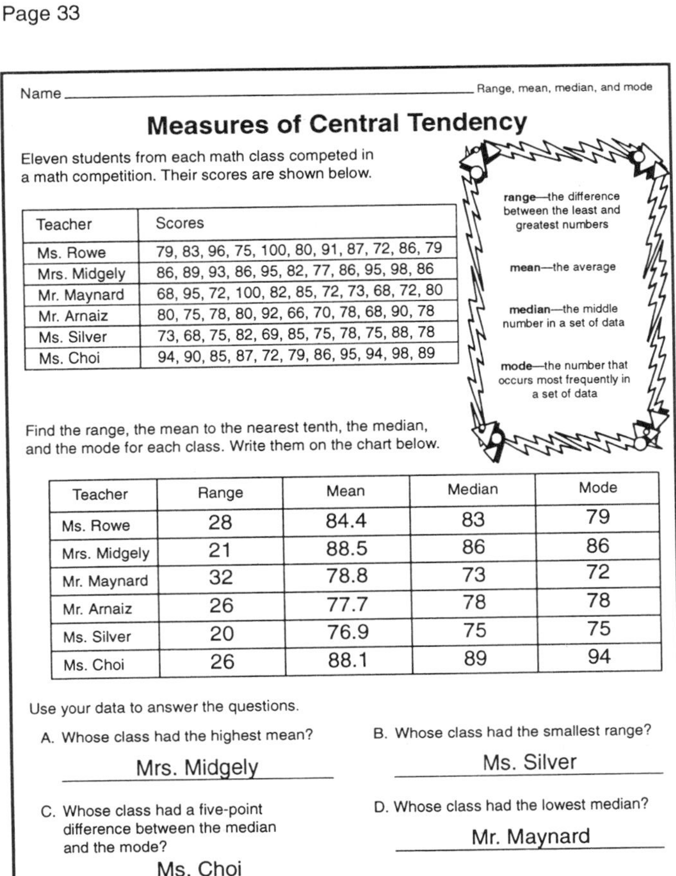

Name ______________________ Range, mean, median, and mode

Measures of Central Tendency

Eleven students from each math class competed in a math competition. Their scores are shown below.

Teacher	Scores
Ms. Rowe	79, 83, 96, 75, 100, 80, 91, 87, 72, 86, 79
Mrs. Midgely	86, 89, 93, 86, 95, 82, 77, 86, 95, 98, 86
Mr. Maynard	68, 95, 72, 100, 82, 85, 72, 73, 68, 72, 80
Mr. Arnaiz	80, 75, 78, 80, 92, 66, 70, 78, 68, 90, 78
Ms. Silver	73, 68, 75, 82, 69, 85, 75, 78, 75, 88, 78
Ms. Choi	94, 90, 85, 87, 72, 79, 86, 95, 94, 98, 89

range—the difference between the least and greatest numbers

mean—the average

median—the middle number in a set of data

mode—the number that occurs most frequently in a set of data

Find the range, the mean to the nearest tenth, the median, and the mode for each class. Write them on the chart below.

Teacher	Range	Mean	Median	Mode
Ms. Rowe	28	84.4	83	79
Mrs. Midgely	21	88.5	86	86
Mr. Maynard	32	78.8	73	72
Mr. Arnaiz	26	77.7	78	78
Ms. Silver	20	76.9	75	75
Ms. Choi	26	88.1	89	94

Use your data to answer the questions.

A. Whose class had the highest mean?
Mrs. Midgely

B. Whose class had the smallest range?
Ms. Silver

C. Whose class had a five-point difference between the median and the mode?
Ms. Choi

D. Whose class had the lowest median?
Mr. Maynard

Page 35

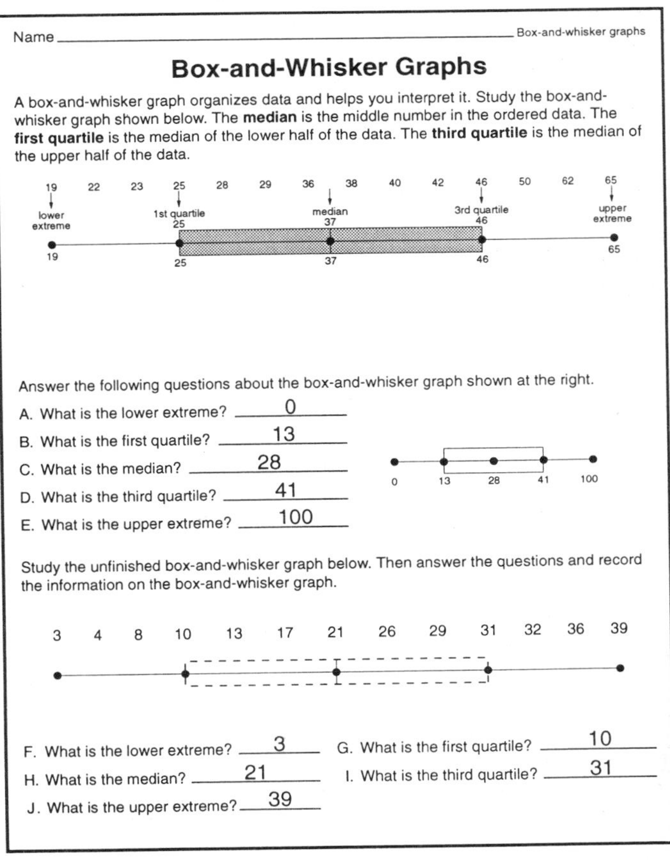

Name ______________________ Box-and-whisker graphs

Box-and-Whisker Graphs

A box-and-whisker graph organizes data and helps you interpret it. Study the box-and-whisker graph shown below. The **median** is the middle number in the ordered data. The **first quartile** is the median of the lower half of the data. The **third quartile** is the median of the upper half of the data.

Answer the following questions about the box-and-whisker graph shown at the right.

A. What is the lower extreme? 0

B. What is the first quartile? 13

C. What is the median? 28

D. What is the third quartile? 41

E. What is the upper extreme? 100

Study the unfinished box-and-whisker graph below. Then answer the questions and record the information on the box-and-whisker graph.

F. What is the lower extreme? 3

G. What is the first quartile? 10

H. What is the median? 21

I. What is the third quartile? 31

J. What is the upper extreme? 39

Page 36

Answer Key

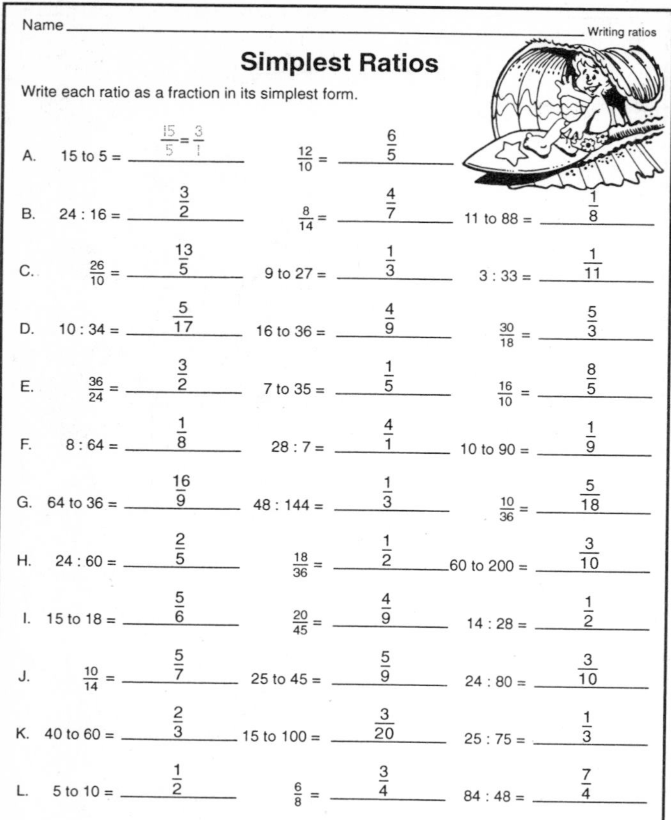

Name ______ Writing ratios

Simplest Ratios

Write each ratio as a fraction in its simplest form.

A.	15 to 5 = $\frac{15}{5} = \frac{3}{1}$	$\frac{12}{10} = \frac{6}{5}$	
B.	24 : 16 = $\frac{3}{2}$	$\frac{8}{14} = \frac{4}{7}$	11 to 88 = $\frac{1}{8}$
C.	$\frac{26}{10} = \frac{13}{5}$	9 to 27 = $\frac{1}{3}$	3 : 33 = $\frac{1}{11}$
D.	10 : 34 = $\frac{5}{17}$	16 to 36 = $\frac{4}{9}$	$\frac{30}{18} = \frac{5}{3}$
E.	$\frac{36}{24} = \frac{3}{2}$	7 to 35 = $\frac{1}{5}$	$\frac{16}{10} = \frac{8}{5}$
F.	8 : 64 = $\frac{1}{8}$	28 : 7 = $\frac{4}{1}$	10 to 90 = $\frac{1}{9}$
G.	64 to 36 = $\frac{16}{9}$	48 : 144 = $\frac{1}{3}$	$\frac{10}{36} = \frac{5}{18}$
H.	24 : 60 = $\frac{2}{5}$	$\frac{18}{36} = \frac{1}{2}$	60 to 200 = $\frac{3}{10}$
I.	15 to 18 = $\frac{5}{6}$	$\frac{20}{45} = \frac{4}{9}$	14 : 28 = $\frac{1}{2}$
J.	$\frac{10}{14} = \frac{5}{7}$	25 to 45 = $\frac{5}{9}$	24 : 80 = $\frac{3}{10}$
K.	40 to 60 = $\frac{2}{3}$	15 to 100 = $\frac{3}{20}$	25 : 75 = $\frac{1}{3}$
L.	5 to 10 = $\frac{1}{2}$	$\frac{6}{8} = \frac{3}{4}$	84 : 48 = $\frac{7}{4}$

Page 37

Name ______ Rates

Rate Per Unit

A **rate** is a ratio that compares quantities of different units. A **unit rate** is a ratio that has 1 as the second term. Write the unit rate for each rate listed below.

A.	48 people in 6 vans: 8 in 1	150 kilometers in 3 hours: 50 in 1	45 cookies in 9 packages: 5 in 1
B.	30 pounds in 6 weeks: 5 in 1	1,250 words in 5 minutes: 250 in 1	28 days in 4 weeks: 7 in 1
C.	90 people in 15 cars: 6 in 1	100 cards in 4 packages: 25 in 1	78 centimeters in 3 seconds: 26 in 1
D.	48 months in 4 years: 12 in 1	144 markers in 6 boxes: 24 in 1	143 players on 11 teams: 13 in 1
E.	550 miles in 10 hours: 55 in 1	720 minutes in 4 trips: 180 in 1	498 milliliters in 2 glasses: 249 in 1
F.	258 kilometers in 3 hours: 86 in 1	300 people in 6 buses: 50 in 1	600 flowers in 150 corsages: 4 in 1
G.	160 pages in 4 hours: 40 in 1	266 rides in 7 weeks: 38 in 1	78 miles in 2 hours: 39 in 1
H.	210 sit-ups in 6 days: 35 in 1	294 minutes for 7 lessons: 42 in 1	40 apples for 5 children: 8 in 1
I.	960 miles in 8 hours: 120 in 1	152 crayons in 8 packages: 19 in 1	560 calories in 5 apples: 112 in 1

Page 38

Name ______ Identifying proportions

Is It a Proportion?

A **proportion** is an equation that has two equivalent ratios. Look at each pair of ratios below. Circle **Y** (yes) if they form a proportion. Circle **N** (no) if they do not.

A.	$\frac{7}{3}$ and $\frac{14}{6}$	(Y)	$\frac{2}{3}$ and $\frac{10}{12}$	(N)	$\frac{8}{4}$ and $\frac{6}{2}$	(N)
B.	$\frac{2}{3}$ and $\frac{8}{12}$	(Y)	$\frac{2}{5}$ and $\frac{3}{10}$	(N)	$\frac{2}{7}$ and $\frac{4}{14}$	(Y)
C.	$\frac{8}{12}$ and $\frac{12}{15}$	(N)	$\frac{8}{3}$ and $\frac{16}{6}$	(Y)	$\frac{28}{21}$ and $\frac{8}{6}$	(N)
D.	$\frac{2}{3}$ and $\frac{4}{9}$	(N)	$\frac{3}{5}$ and $\frac{9}{15}$	(Y)	$\frac{4}{7}$ and $\frac{8}{14}$	(Y)
E.	$\frac{4}{10}$ and $\frac{6}{15}$	(Y)	$\frac{7}{12}$ and $\frac{3}{4}$	(N)	$\frac{12}{54}$ and $\frac{2}{9}$	(Y)
F.	$\frac{5}{7}$ and $\frac{35}{49}$	(Y)	$\frac{4}{14}$ and $\frac{6}{21}$	(Y)	$\frac{12}{25}$ and $\frac{60}{125}$	(Y)
G.	$\frac{7}{9}$ and $\frac{63}{88}$	(N)	$\frac{9}{13}$ and $\frac{45}{72}$	(N)	$\frac{6}{15}$ and $\frac{45}{150}$	(N)
H.	$\frac{10}{5}$ and $\frac{15}{10}$	(N)	$\frac{4}{9}$ and $\frac{16}{36}$	(Y)	$\frac{8}{2}$ and $\frac{40}{10}$	(Y)
I.	$\frac{1}{7}$ and $\frac{10}{70}$	(Y)	$\frac{43}{85}$ and $\frac{86}{160}$	(N)	$\frac{2}{9}$ and $\frac{27}{4}$	(N)
J.	$\frac{15}{9}$ and $\frac{75}{45}$	(Y)	$\frac{63}{9}$ and $\frac{9}{1}$	(Y)	$\frac{45}{50}$ and $\frac{8}{10}$	(N)
K.	$\frac{46}{86}$ and $\frac{44}{84}$	(N)	$\frac{51}{52}$ and $\frac{49}{50}$	(N)		
L.	$\frac{4}{7}$ and $\frac{5}{9}$	(N)	$\frac{2}{18}$ and $\frac{20}{180}$	(Y)		

Page 39

Name ______ Solving proportions

Solving Proportions

You can cross multiply to find the missing number in a proportion. Use mental math or a calculator to find the missing numbers in the proportions listed below.

Mental Math	Calculator Math
$\frac{3}{n} = \frac{9}{21}$ 3 x 3 = 9 so n x 3 = 21 n must be 7	$\frac{3}{n} = \frac{9}{21}$ Cross multiply. 3 x 21 = 9 x n 63 = 9 x n n must be 7

A.	$\frac{3}{8} = \frac{12}{n}$, n = 32	$\frac{5}{6} = \frac{n}{12}$, n = 10		
B.	$\frac{n}{6} = \frac{33}{18}$, n = 11	$\frac{16}{72} = \frac{n}{9}$, n = 2	$\frac{n}{40} = \frac{6}{10}$, n = 24	$\frac{n}{5} = \frac{16}{20}$, n = 4
C.	$\frac{8}{12} = \frac{6}{n}$, n = 9	$\frac{3}{9} = \frac{n}{15}$, n = 5	$\frac{15}{5} = \frac{12}{n}$, n = 4	$\frac{1}{2} = \frac{n}{4.2}$, n = 2.1
D.	$\frac{15}{20} = \frac{n}{4}$, n = 3	$\frac{1}{12} = \frac{1.5}{n}$, n = 18	$\frac{4}{5} = \frac{36}{n}$, n = 45	$\frac{8}{5} = \frac{6}{n}$, n = 3.75
E.	$\frac{2}{n} = \frac{5}{7.5}$, n = 3	$\frac{14}{8} = \frac{n}{20}$, n = 35	$\frac{n}{0.8} = \frac{7}{.8}$, n = 0.7	$\frac{10}{13} = \frac{30}{n}$, n = 39
F.	$\frac{7}{9} = \frac{63}{n}$, n = 81	$\frac{n}{3} = \frac{15}{18}$, n = 2.5	$\frac{15}{12} = \frac{n}{4}$, n = 5	$\frac{7}{8} = \frac{21}{n}$, n = 24

Page 40

Answer Key

Name ____________ Finding unit prices

Unit Pricing

Find the unit prices. Round to the nearest cent if necessary.

A.	3 for \$0.96 — \$ 0.32	7 for \$1.61 — \$ 0.23	
B.	9 for \$0.54 — \$ 0.06	4 for \$1.76 — \$ 0.44	5 for \$1.55 — \$ 0.31
C.	5 for \$19.95 — \$ 3.99	6 for \$1.41 — \$ 0.24	$1\frac{1}{2}$ for \$3.45 — \$ 2.30
D.	12 for \$13.32 — \$ 1.11	4 for \$5.00 — \$ 1.25	3 for \$2.91 — \$ 0.97
E.	3 for \$228.00 — \$ 76.00	20 for \$45.00 — \$ 2.25	8 for \$4.68 — \$ 0.59
F.	$2\frac{1}{4}$ for \$1.00 — \$ 0.44	8 for \$10.00 — \$ 1.25	5 for \$67.00 — \$ 13.40
G.	5 for \$18.75 — \$ 3.75	11 for \$1.21 — \$ 0.11	16 for \$8.80 — \$ 0.55
H.	3 for \$4.74 — \$ 1.58	4 for \$3.00 — \$ 0.75	18 for \$50.00 — \$ 2.78
I.	25 for \$20.00 — \$ 0.80	5 for \$49.75 — \$ 9.95	6 for \$0.88 — \$ 0.15
J.	3 for \$0.69 — \$ 0.23	$4\frac{1}{2}$ for \$90.00 — \$ 20.00	23 for \$57.50 — \$ 2.50

Page 41

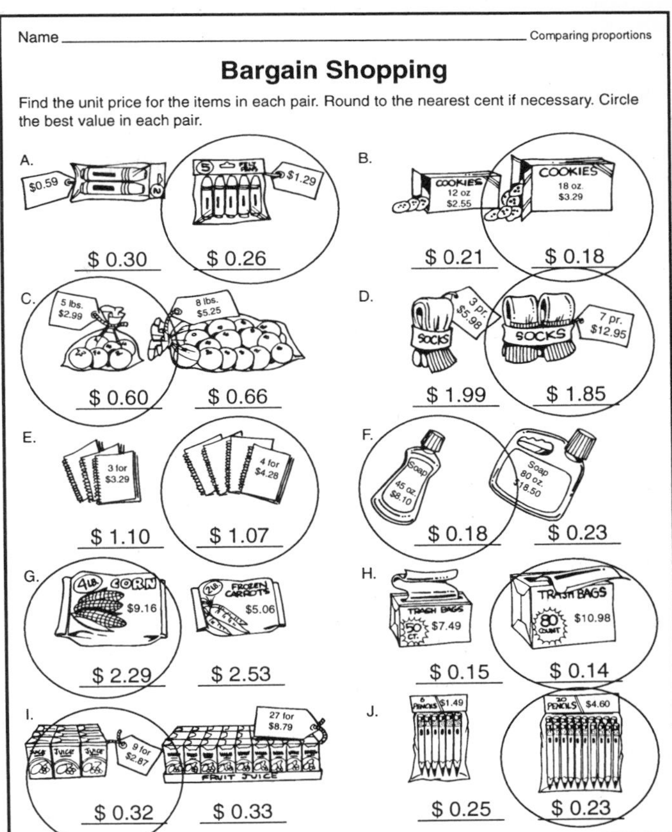

Name ____________ Comparing proportions

Bargain Shopping

Find the unit price for the items in each pair. Round to the nearest cent if necessary. Circle the best value in each pair.

A.	\$ 0.30	\$ 0.26	B.	\$ 0.21	\$ 0.18
C.	\$ 0.60	\$ 0.66	D.	\$ 1.99	\$ 1.85
E.	\$ 1.10	\$ 1.07	F.	\$ 0.18	\$ 0.23
G.	\$ 2.29	\$ 2.53	H.	\$ 0.15	\$ 0.14
I.	\$ 0.32	\$ 0.33	J.	\$ 0.25	\$ 0.23

Page 42

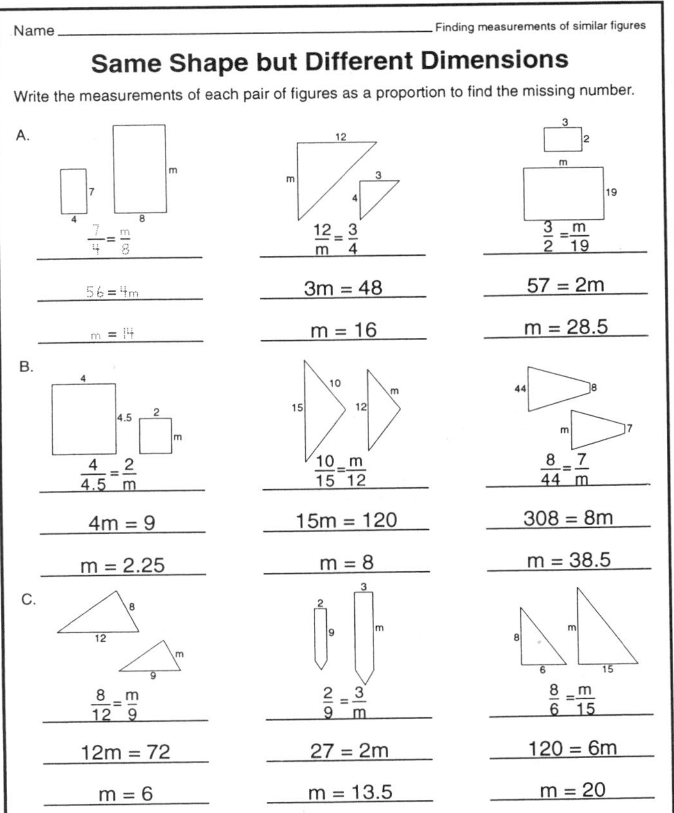

Name ____________ Finding measurements of similar figures

Same Shape but Different Dimensions

Write the measurements of each pair of figures as a proportion to find the missing number.

A.	$\frac{7}{4} = \frac{m}{8}$; $56 = 4m$; $m = 14$	$\frac{12}{m} = \frac{3}{4}$; $3m = 48$; $m = 16$	$\frac{3}{2} = \frac{m}{19}$; $57 = 2m$; $m = 28.5$
B.	$\frac{4}{4.5} = \frac{2}{m}$; $4m = 9$; $m = 2.25$	$\frac{10}{15} = \frac{m}{12}$; $15m = 120$; $m = 8$	$\frac{8}{44} = \frac{7}{m}$; $308 = 8m$; $m = 38.5$
C.	$\frac{8}{12} = \frac{m}{9}$; $12m = 72$; $m = 6$	$\frac{2}{9} = \frac{3}{m}$; $27 = 2m$; $m = 13.5$	$\frac{8}{6} = \frac{m}{15}$; $120 = 6m$; $m = 20$

Page 43

Name ____________ Writing percents

Writing Percents

Write each decimal or fraction as a percent.

Percent means "out of 100."

A.	$\frac{35}{100}$ = 35%	$\frac{17}{100}$ = 17%	0.93 = 93%	
B.	0.41 = 41%	0.45 = 45%	0.91 = 91%	
C.	0.23 = 23%	$\frac{75}{100}$ = 75%	0.72 = 72%	0.51 = 51%
D.	$\frac{81}{100}$ = 81%	0.01 = 1%	$\frac{25}{100}$ = 25%	$\frac{73}{100}$ = 73%
E.	$\frac{27}{100}$ = 27%	$\frac{1}{2}$ = 50%	$\frac{43}{100}$ = 43%	0.03 = 3%
F.	0.31 = 31%	0.1 = 10%	$\frac{4}{10}$ = 40%	0.85 = 85%
G.	0.05 = 5%	$\frac{39}{100}$ = 39%	$\frac{6}{10}$ = 60%	$\frac{29}{100}$ = 29%
H.	$\frac{9}{100}$ = 9%	$\frac{8}{10}$ = 80%	0.08 = 8%	0.15 = 15%
I.	0.63 = 63%	0.2 = 20%	$\frac{45}{100}$ = 45%	0.4 = 40%
J.	0.71 = 71%	0.86 = 86%	$\frac{23}{100}$ = 23%	0.07 = 7%
K.	$\frac{97}{100}$ = 97%	$\frac{7}{100}$ = 7%	$\frac{7}{10}$ = 70%	0.98 = 98%
L.	0.42 = 42%	$\frac{9}{10}$ = 90%	$\frac{33}{100}$ = 33%	0.66 = 66%
M.	0.31 = 31%	$\frac{99}{100}$ = 99%	$\frac{47}{100}$ = 47%	0.02 = 2%

Page 44

Answer Key

Name________ Percents and decimals

Making Decimals From Percents

To change a percent to a decimal, move the decimal point two digits to the left.
Change each percent below to a decimal.

46% = 0.46

A.	95% = 0.95	82% = 0.82	43% = 0.43
B.	17% = 0.17	68% = 0.68	48% = 0.48
C.	71% = 0.71	4% = 0.04	73% = 0.73
D.	55% = 0.55	30% = 0.3	10% = 0.1
E.	7% = 0.07	15% = 0.15	9% = 0.09
F.	84% = 0.84	52% = 0.52	32% = 0.32
G.	94% = 0.94	2% = 0.02	11% = 0.11
H.	50% = 0.50	34% = 0.34	5% = 0.05
I.	19% = 0.19	100% = 1	26% = 0.26
J.	20% = 0.2	1% = 0.01	16% = 0.16
K.	12% = 0.12	21% = 0.21	51% = 0.51
L.	6% = 0.06	18% = 0.18	3% = 0.03
M.	53% = 0.53	8% = 0.08	13% = 0.13
N.	88% = 0.88	14% = 0.14	99% = 0.99

Page 45

Name________ Percents and fractions

Making Fractions From Percents

Write each percent below as a fraction with a denominator of 100. Then write the fraction in its simplest form.

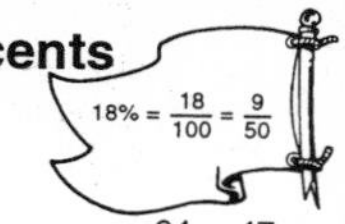

$18\% = \frac{18}{100} = \frac{9}{50}$

A.	$49\% = \frac{49}{100}$	$75\% = \frac{75}{100} = \frac{3}{4}$	$94\% = \frac{94}{100} = \frac{47}{50}$
B.	$65\% = \frac{65}{100} = \frac{13}{20}$	$17\% = \frac{17}{100}$	$2\% = \frac{2}{100} = \frac{1}{50}$
C.	$100\% = \frac{100}{100} = 1$	$6\% = \frac{6}{100} = \frac{3}{50}$	$56\% = \frac{56}{100} = \frac{14}{25}$
D.	$8\% = \frac{8}{100} = \frac{2}{25}$	$80\% = \frac{80}{100} = \frac{4}{5}$	$99\% = \frac{99}{100}$
E.	$30\% = \frac{30}{100} = \frac{3}{10}$	$28\% = \frac{28}{100} = \frac{7}{25}$	$73\% = \frac{73}{100}$
F.	$25\% = \frac{25}{100} = \frac{1}{4}$	$93\% = \frac{93}{100}$	$10\% = \frac{10}{100} = \frac{1}{10}$
G.	$5\% = \frac{5}{100} = \frac{1}{20}$	$4\% = \frac{4}{100} = \frac{1}{25}$	$95\% = \frac{95}{100} = \frac{19}{20}$
H.	$7\% = \frac{7}{100}$	$40\% = \frac{40}{100} = \frac{2}{5}$	$50\% = \frac{50}{100} = \frac{1}{2}$
I.	$90\% = \frac{90}{100} = \frac{9}{10}$	$3\% = \frac{3}{100}$	$62\% = \frac{62}{100} = \frac{31}{50}$
J.	$60\% = \frac{60}{100} = \frac{3}{5}$	$85\% = \frac{85}{100} = \frac{17}{20}$	$48\% = \frac{48}{100} = \frac{12}{25}$
K.	$1\% = \frac{1}{100}$	$46\% = \frac{46}{100} = \frac{23}{50}$	$45\% = \frac{45}{100} = \frac{9}{20}$
L.	$12\% = \frac{12}{100} = \frac{3}{25}$	$74\% = \frac{74}{100} = \frac{37}{50}$	$20\% = \frac{20}{100} = \frac{1}{5}$
M.	$15\% = \frac{15}{100} = \frac{3}{20}$	$88\% = \frac{88}{100} = \frac{22}{25}$	$66\% = \frac{66}{100} = \frac{33}{50}$

Page 46

Name________ Percent

Special Percents

Write each percent in decimal form and as a fraction or mixed number.

Percent	Decimal	Fraction or Mixed Number	Percent	Decimal	Fraction or Mixed Number
A. 150%	1.5	$1\frac{1}{2}$	M. 227%	2.27	$2\frac{27}{100}$
B. 120%	1.2	$1\frac{1}{5}$	N. 250%	2.5	$2\frac{1}{2}$
C. 66.66 %	0.6666	$\frac{3,333}{5,000}$	O. 2.4%	0.024	$\frac{3}{125}$
D. 0.5 %	0.005	$\frac{1}{200}$	P. 312.6%	3.126	$3\frac{63}{500}$
E. 6.9%	0.069	$\frac{69}{1,000}$	Q. 33.33%	0.3333	$\frac{3,333}{10,000}$
F. 87.5%	0.875	$\frac{7}{8}$	R. 37.5%	0.375	$\frac{3}{8}$
G. 600%	6.00	6	S. 12.5%	0.125	$\frac{1}{8}$
H. 2.5%	0.025	$\frac{1}{40}$	T. 0.45%	0.0045	$\frac{9}{2,000}$
I. 0.9%	0.009	$\frac{9}{1,000}$	U. 410.7%	4.107	$4\frac{107}{1,000}$
J. 123%	1.23	$1\frac{23}{100}$	V. 12.3%	0.123	$\frac{123}{1,000}$
K. 1.23%	0.0123	$\frac{123}{10,000}$	W. 90%	0.90	$\frac{9}{10}$
L. 0.03%	0.0003	$\frac{3}{10,000}$	X. 0.25%	0.0025	$\frac{1}{400}$

Page 47

Name________ Percent of a number

Percent of a Number

Find the percent of each number by using a calculator or by changing the percents to fractions or decimals and multiplying.

75% of 96
$\frac{3}{4} \times \frac{96}{1} = 72$

	A. 50% of 108 = 54	30% of 50 = 15
B. 6% of 20 = 1.2	85% of 150 = 127.5	42% of 147 = 61.74
C. 80% of 65 = 52	30% of 30 = 9	8% of 40 = 3.2
D. 150% of 50 = 75	25% of 60 = 15	300% of 85 = 255
E. 28% of 400 = 112	75% of 48 = 36	25% of 200 = 50
F. 35% of 50 = 17.5	39% of 500 = 195	60% of 90 = 54
G. 25% of 124 = 31	$33\frac{1}{3}$% of 96 = 32	26% of 130 = 33.8
H. 13% of 600 = 78	70% of 148 = 103.6	2% of 63 = 1.26

Page 48

Answer Key

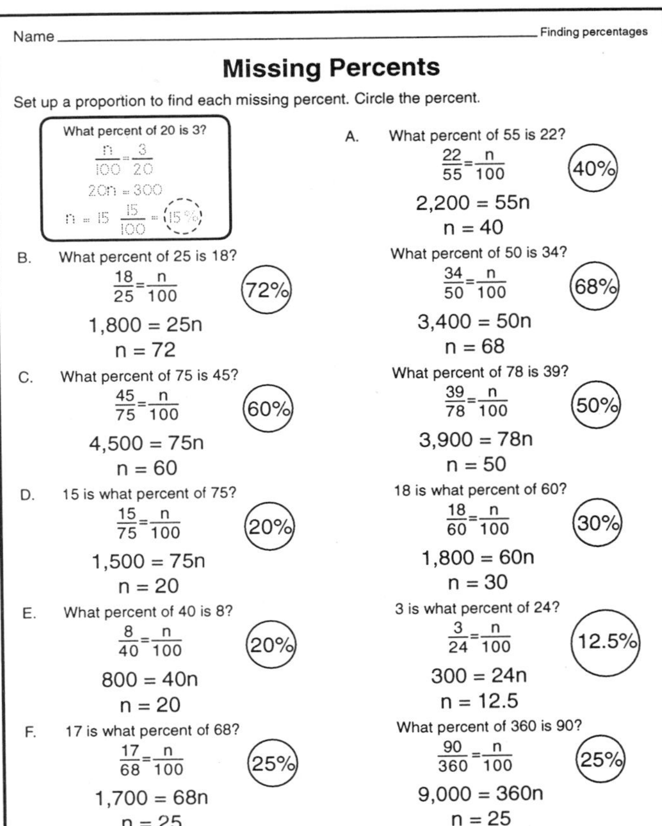

Name ______ Finding percentages

Missing Percents

Set up a proportion to find each missing percent. Circle the percent.

What percent of 20 is 3?
$\frac{n}{100} = \frac{3}{20}$
$20n = 300$
$n = 15$ $\frac{15}{100} = (15\%)$

A. What percent of 55 is 22?
$\frac{22}{55} = \frac{n}{100}$ (40%)
$2{,}200 = 55n$
$n = 40$

B. What percent of 25 is 18?
$\frac{18}{25} = \frac{n}{100}$ (72%)
$1{,}800 = 25n$
$n = 72$

What percent of 50 is 34?
$\frac{34}{50} = \frac{n}{100}$ (68%)
$3{,}400 = 50n$
$n = 68$

C. What percent of 75 is 45?
$\frac{45}{75} = \frac{n}{100}$ (60%)
$4{,}500 = 75n$
$n = 60$

What percent of 78 is 39?
$\frac{39}{78} = \frac{n}{100}$ (50%)
$3{,}900 = 78n$
$n = 50$

D. 15 is what percent of 75?
$\frac{15}{75} = \frac{n}{100}$ (20%)
$1{,}500 = 75n$
$n = 20$

18 is what percent of 60?
$\frac{18}{60} = \frac{n}{100}$ (30%)
$1{,}800 = 60n$
$n = 30$

E. What percent of 40 is 8?
$\frac{8}{40} = \frac{n}{100}$ (20%)
$800 = 40n$
$n = 20$

3 is what percent of 24?
$\frac{3}{24} = \frac{n}{100}$ (12.5%)
$300 = 24n$
$n = 12.5$

F. 17 is what percent of 68?
$\frac{17}{68} = \frac{n}{100}$ (25%)
$1{,}700 = 68n$
$n = 25$

What percent of 360 is 90?
$\frac{90}{360} = \frac{n}{100}$ (25%)
$9{,}000 = 360n$
$n = 25$

Page 49

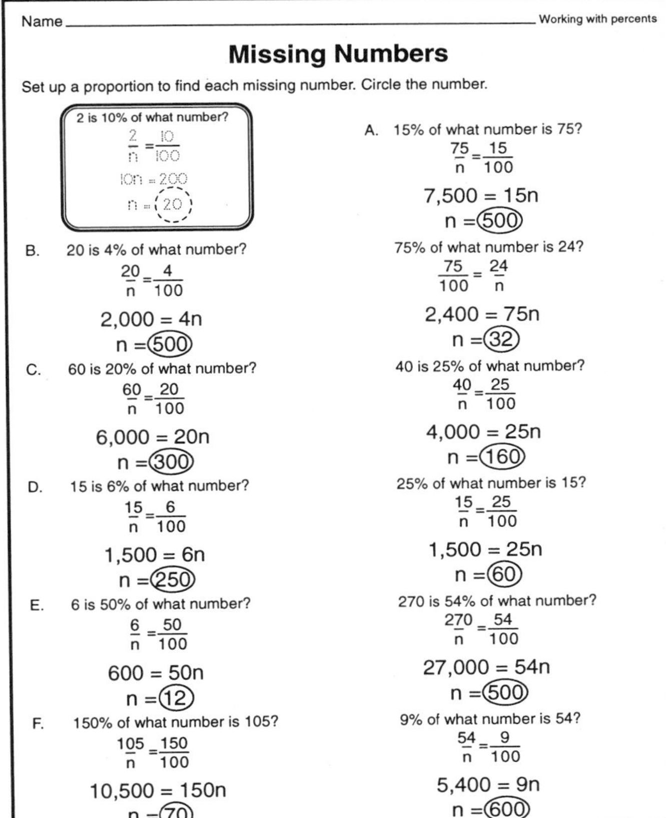

Name ______ Working with percents

Missing Numbers

Set up a proportion to find each missing number. Circle the number.

2 is 10% of what number?
$\frac{2}{n} = \frac{10}{100}$
$10n = 200$
$n = (20)$

A. 15% of what number is 75?
$\frac{75}{n} = \frac{15}{100}$
$7{,}500 = 15n$
$n = (500)$

B. 20 is 4% of what number?
$\frac{20}{n} = \frac{4}{100}$
$2{,}000 = 4n$
$n = (500)$

75% of what number is 24?
$\frac{75}{100} = \frac{24}{n}$
$2{,}400 = 75n$
$n = (32)$

C. 60 is 20% of what number?
$\frac{60}{n} = \frac{20}{100}$
$6{,}000 = 20n$
$n = (300)$

40 is 25% of what number?
$\frac{40}{n} = \frac{25}{100}$
$4{,}000 = 25n$
$n = (160)$

D. 15 is 6% of what number?
$\frac{15}{n} = \frac{6}{100}$
$1{,}500 = 6n$
$n = (250)$

25% of what number is 15?
$\frac{15}{n} = \frac{25}{100}$
$1{,}500 = 25n$
$n = (60)$

E. 6 is 50% of what number?
$\frac{6}{n} = \frac{50}{100}$
$600 = 50n$
$n = (12)$

270 is 54% of what number?
$\frac{270}{n} = \frac{54}{100}$
$27{,}000 = 54n$
$n = (500)$

F. 150% of what number is 105?
$\frac{105}{n} = \frac{150}{100}$
$10{,}500 = 150n$
$n = (70)$

9% of what number is 54?
$\frac{54}{n} = \frac{9}{100}$
$5{,}400 = 9n$
$n = (600)$

Page 50

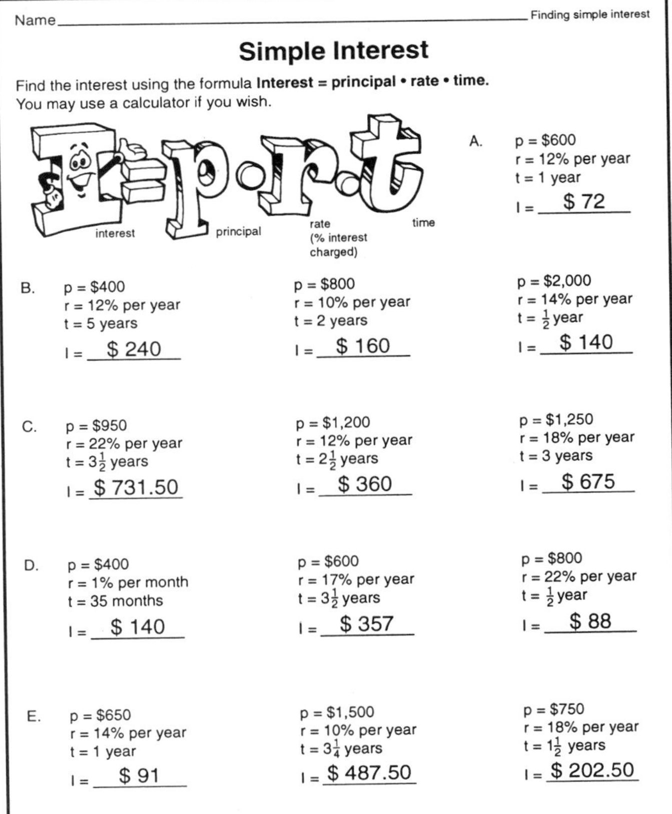

Name ______ Finding simple interest

Simple Interest

Find the interest using the formula **Interest = principal • rate • time.** You may use a calculator if you wish.

I = p • r • t
interest
principal
rate (% interest charged)
time

A. p = \$600
r = 12% per year
t = 1 year
I = \$ 72

B. p = \$400
r = 12% per year
t = 5 years
I = \$ 240

p = \$800
r = 10% per year
t = 2 years
I = \$ 160

p = \$2,000
r = 14% per year
t = $\frac{1}{2}$ year
I = \$ 140

C. p = \$950
r = 22% per year
t = $3\frac{1}{2}$ years
I = \$ 731.50

p = \$1,200
r = 12% per year
t = $2\frac{1}{2}$ years
I = \$ 360

p = \$1,250
r = 18% per year
t = 3 years
I = \$ 675

D. p = \$400
r = 1% per month
t = 35 months
I = \$ 140

p = \$600
r = 17% per year
t = $3\frac{1}{2}$ years
I = \$ 357

p = \$800
r = 22% per year
t = $\frac{1}{2}$ year
I = \$ 88

E. p = \$650
r = 14% per year
t = 1 year
I = \$ 91

p = \$1,500
r = 10% per year
t = $3\frac{1}{4}$ years
I = \$ 487.50

p = \$750
r = 18% per year
t = $1\frac{1}{2}$ years
I = \$ 202.50

Page 51

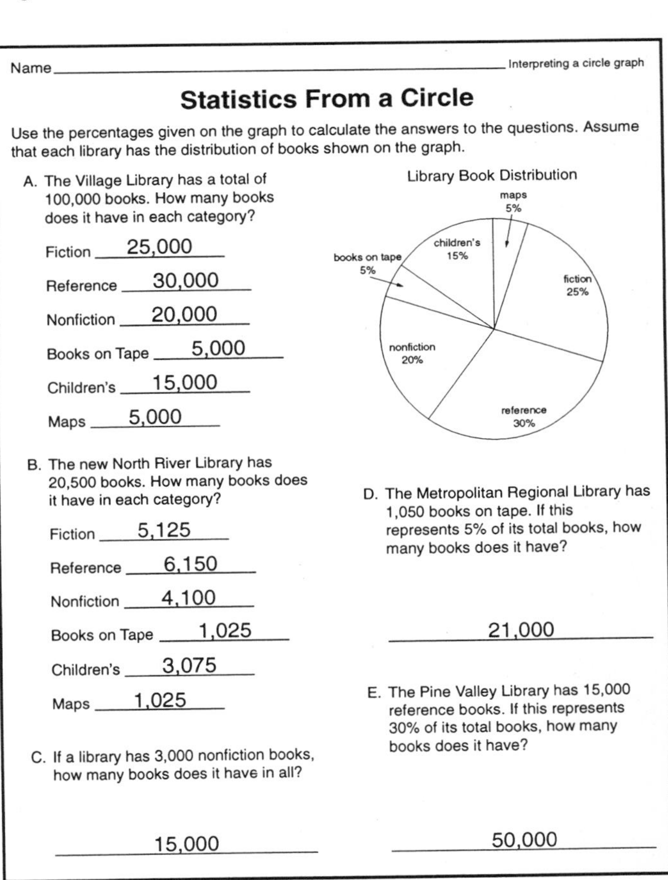

Name ______ Interpreting a circle graph

Statistics From a Circle

Use the percentages given on the graph to calculate the answers to the questions. Assume that each library has the distribution of books shown on the graph.

Library Book Distribution
maps 5%
children's 15%
books on tape 5%
fiction 25%
nonfiction 20%
reference 30%

A. The Village Library has a total of 100,000 books. How many books does it have in each category?

Fiction 25,000
Reference 30,000
Nonfiction 20,000
Books on Tape 5,000
Children's 15,000
Maps 5,000

B. The new North River Library has 20,500 books. How many books does it have in each category?

Fiction 5,125
Reference 6,150
Nonfiction 4,100
Books on Tape 1,025
Children's 3,075
Maps 1,025

C. If a library has 3,000 nonfiction books, how many books does it have in all?

15,000

D. The Metropolitan Regional Library has 1,050 books on tape. If this represents 5% of its total books, how many books does it have?

21,000

E. The Pine Valley Library has 15,000 reference books. If this represents 30% of its total books, how many books does it have?

50,000

Page 52

Answer Key

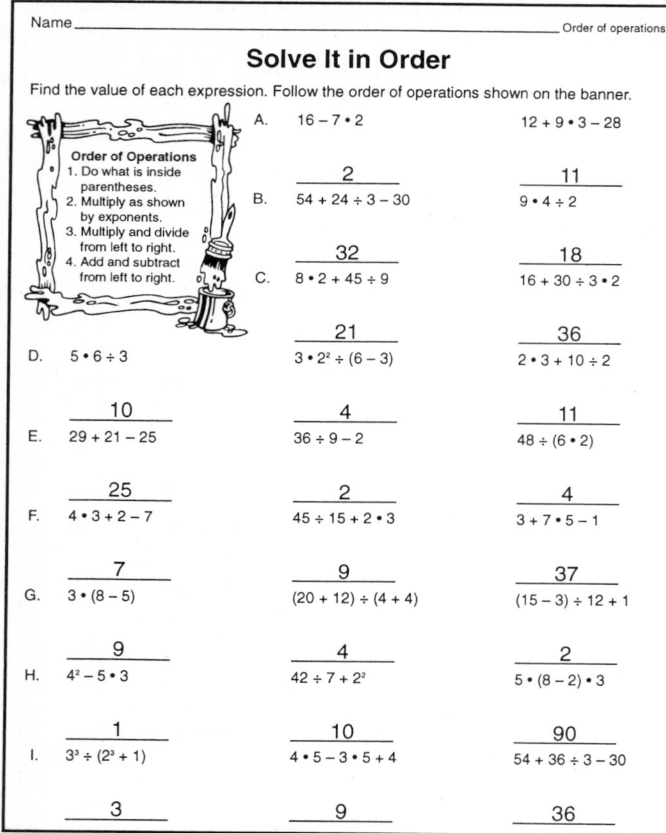

Name ______________________ Order of operations

Solve It in Order

Find the value of each expression. Follow the order of operations shown on the banner.

Order of Operations
1. Do what is inside parentheses.
2. Multiply as shown by exponents.
3. Multiply and divide from left to right.
4. Add and subtract from left to right.

A. 16 − 7 • 2 = **2** | 12 + 9 • 3 − 28 = **11**

B. 54 + 24 ÷ 3 − 30 = **32** | 9 • 4 ÷ 2 = **18**

C. 8 • 2 + 45 ÷ 9 = **21** | 16 + 30 ÷ 3 • 2 = **36**

D. 5 • 6 ÷ 3 = **10** | $3 \cdot 2^2 \div (6 - 3)$ = **4** | 2 • 3 + 10 ÷ 2 = **11**

E. 29 + 21 − 25 = **25** | 36 ÷ 9 − 2 = **2** | 48 ÷ (6 • 2) = **4**

F. 4 • 3 + 2 − 7 = **7** | 45 ÷ 15 + 2 • 3 = **9** | 3 + 7 • 5 − 1 = **37**

G. 3 • (8 − 5) = **9** | (20 + 12) ÷ (4 + 4) = **4** | (15 − 3) ÷ 12 + 1 = **2**

H. $4^2 - 5 \cdot 3$ = **1** | $42 \div 7 + 2^2$ = **10** | 5 • (8 − 2) • 3 = **90**

I. $3^3 \div (2^3 + 1)$ = **3** | 4 • 5 − 3 • 5 + 4 = **9** | 54 + 36 ÷ 3 − 30 = **36**

Page 53

Name ______________________ Translating algebraic expressions

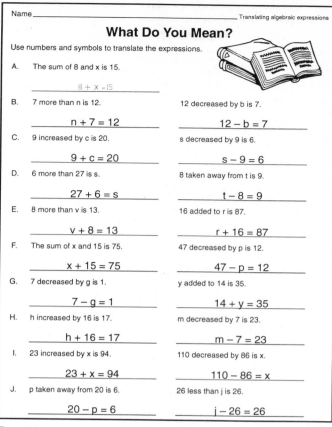

What Do You Mean?

Use numbers and symbols to translate the expressions.

A. The sum of 8 and x is 15. — 8 + x = 15

B. 7 more than n is 12. — n + 7 = 12 | 12 decreased by b is 7. — 12 − b = 7

C. 9 increased by c is 20. — 9 + c = 20 | s decreased by 9 is 6. — s − 9 = 6

D. 6 more than 27 is s. — 27 + 6 = s | 8 taken away from t is 9. — t − 8 = 9

E. 8 more than v is 13. — v + 8 = 13 | 16 added to r is 87. — r + 16 = 87

F. The sum of x and 15 is 75. — x + 15 = 75 | 47 decreased by p is 12. — 47 − p = 12

G. 7 decreased by g is 1. — 7 − g = 1 | y added to 14 is 35. — 14 + y = 35

H. h increased by 16 is 17. — h + 16 = 17 | m decreased by 7 is 23. — m − 7 = 23

I. 23 increased by x is 94. — 23 + x = 94 | 110 decreased by 86 is x. — 110 − 86 = x

J. p taken away from 20 is 6. — 20 − p = 6 | 26 less than j is 26. — j − 26 = 26

Page 54

Name ______________________ Function tables

Number Patterns

Use the rule given to complete each function table.

A.

x	3	5	7	9	11
x + 2	5	7	9	11	13

B.

e	1	5	9	13	17
e + 9	10	14	18	22	26

g	2	4	6	8	10
10 − g	8	6	4	2	0

C.

k	30	45	60	75	90
k − 15	15	30	45	60	75

t	13	28	43	58	73
37 + t	50	65	80	95	110

D.

w	91	82	73	64	55
100 − w	9	18	27	36	45

f	11	22	33	44	55
f + 99	110	121	132	143	154

Write the rule for each function table. Then complete the table.

E.

h	5	6	7	8	9
h + 3	8	9	10	11	12

j	15	20	25	30	35
j + 1	16	21	26	31	36

F.

m	10	12	14	16	18
m − 4	6	8	10	12	14

p	28	35	42	49	56
p − 7	21	28	35	42	49

G.

r	56	65	74	83	92
r − 10	46	55	64	73	82

v	90	80	70	60	50
v − 11	79	69	59	49	39

H.

y	37	35	33	31	29
y − 9	28	26	24	22	20

b	59	62	65	68	71
b + 3	62	65	68	71	74

Page 55

Name ______________________ Evaluating expressions

Find the Values

Evaluate each expression by replacing each variable (letter) with its given value.

Let x = 10.

A. 3.5 + x = 13.5 | 12 − x = 2 | (8 + 9) − x = 7

B. 14 + x = 24 | 43.2 − x = 33.2 | x − 5.3 = 4.7

Let y = 14.5.

C. 9 + y = 23.5 | 56 − y = 41.5 | 14.5 − y = 0

D. 53.5 + y = 68 | y − 9.3 = 5.2 | 15 − y = 0.5

Let m = 25. Let n = 10.

E. 43 + (m + n) = 78 | 50 + (m − n) = 65 | (95 + m) − n = 110

F. (67 − m) + n = 52 | (m − 10) + n = 25 | n − 6 + m = 29

G. 49 − (m + n) = 14 | (m − n) − 15 = 0 | n + (30 − m) = 15

Let r = 9. Let s = 8. Let t = 5.

H. (s + s) − 5 = 11 | (9 + 57) + (r − t) = 70 | (r − s) + t = 6

I. t + (r − s) = 6 | (12 + t) − (r + s) = 0 | (r − s) + 9 = 10

J. 16 + (t + r) = 30 | 100 + t − r = 96 | 50 + (r − t) = 54

K. (r + s) − 16 = 1 | 87 − (r + s + t) = 65 | 15 − (r + t) = 1

Page 56

Answer Key

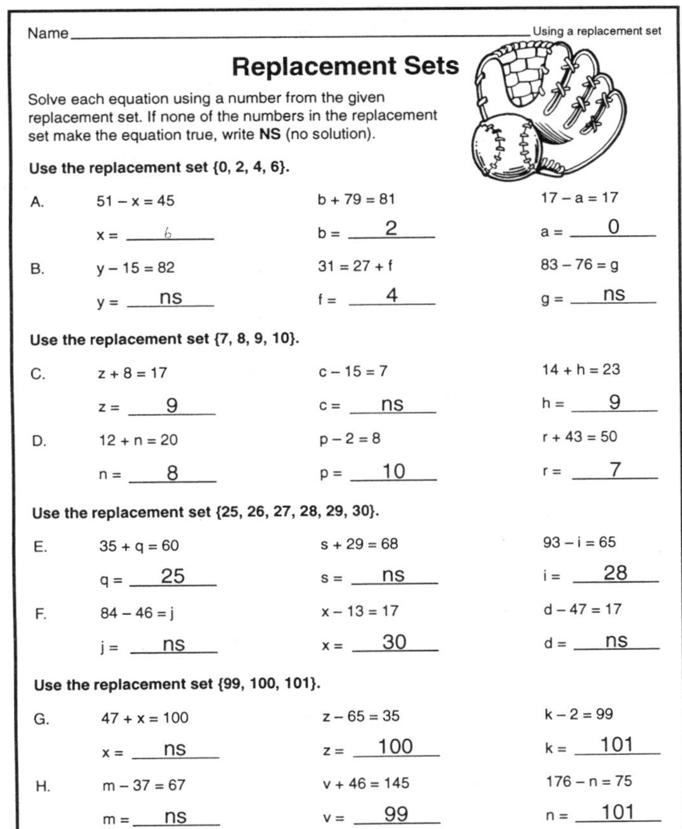

Name________________ Using a replacement set

Replacement Sets

Solve each equation using a number from the given replacement set. If none of the numbers in the replacement set make the equation true, write **NS** (no solution).

Use the replacement set {0, 2, 4, 6}.

A. $51 - x = 45$ $x =$ 6 $b + 79 = 81$ $b =$ 2 $17 - a = 17$ $a =$ 0

B. $y - 15 = 82$ $y =$ ns $31 = 27 + f$ $f =$ 4 $83 - 76 = g$ $g =$ ns

Use the replacement set {7, 8, 9, 10}.

C. $z + 8 = 17$ $z =$ 9 $c - 15 = 7$ $c =$ ns $14 + h = 23$ $h =$ 9

D. $12 + n = 20$ $n =$ 8 $p - 2 = 8$ $p =$ 10 $r + 43 = 50$ $r =$ 7

Use the replacement set {25, 26, 27, 28, 29, 30}.

E. $35 + q = 60$ $q =$ 25 $s + 29 = 68$ $s =$ ns $93 - i = 65$ $i =$ 28

F. $84 - 46 = j$ $j =$ ns $x - 13 = 17$ $x =$ 30 $d - 47 = 17$ $d =$ ns

Use the replacement set {99, 100, 101}.

G. $47 + x = 100$ $x =$ ns $z - 65 = 35$ $z =$ 100 $k - 2 = 99$ $k =$ 101

H. $m - 37 = 67$ $m =$ ns $v + 46 = 145$ $v =$ 99 $176 - n = 75$ $n =$ 101

Page 57

Name________________ Solving addition equations

Addition Equations

$x + 9 = 28$
$x + 9 - \mathbf{9} = 28 - \mathbf{9}$
$x = 19$

To solve an addition equation, subtract the same number from both sides to make the variable stand alone.

A. $x + 8 = 13$; $x + 8 - 8 = 13 - 8$; $x = 5$ $4 + x = 12$; $4 + x - 4 = 12 - 4$; $x = 8$ $x + 6 = 17$; $x + 6 - 6 = 17 - 6$; $x = 11$

B. $x + 5 = 15$; $x + 5 - 5 = 15 - 5$; $x = 10$ $x + 21 = 34$; $x + 21 - 21 = 34 - 21$; $x = 13$ $x + 38 = 81$; $x + 38 - 38 = 81 - 38$; $x = 43$

C. $x + 29 = 65$; $x + 29 - 29 = 65 - 29$; $x = 36$ $46 + x = 95$; $46 - 46 + x = 95 - 46$; $x = 49$ $x + 6.6 = 7.2$; $x + 6.6 - 6.6 = 7.2 - 6.6$; $x = 0.6$

D. $x + 19 = 48$; $x + 19 - 19 = 48 - 19$; $x = 29$ $3.6 + x = 4.9$; $3.6 - 3.6 + x = 4.9 - 3.6$; $x = 1.3$ $815 + x = 902$; $815 - 815 + x = 902 - 815$; $x = 87$

E. $x + 13.8 = 15.6$; $x + 13.8 - 13.8 = 15.6 - 13.8$; $x = 1.8$ $347 + x = 409$; $347 - 347 + x = 409 - 347$; $x = 62$ $x + 12 = 30.1$; $x + 12 - 12 + 30.1 - 12$; $x = 18.1$

F. $169 + x = 490$; $169 - 169 + x = 490 - 169$; $x = 321$ $x + 6.7 = 24.5$; $x + 6.7 - 6.7 = 24.5 - 6.7$; $x = 17.8$ $196 + x = 500$; $196 + x - 196 = 500 - 196$; $x = 304$

Page 58

Name________________ Solving subtraction equations

Subtraction Equations

$y - 21 = 32$
$y - 21 + \mathbf{21} = 32 + \mathbf{21}$
$y = 53$

To solve a subtraction equation, add the same number to both sides to make the variable stand alone.

A. $y - 12 = 25$; $y - 12 + 12 = 25 + 12$; $y = 37$ $y - 19 = 21$; $y - 19 + 19 = 21 + 19$; $y = 40$ $y - 68 = 229$; $y - 68 + 68 = 229 + 68$; $y = 297$

B. $y - 56 = 7$; $y - 56 + 56 = 7 + 56$; $y = 63$ $y - 42 = 67$; $y - 42 + 42 = 67 + 42$; $y = 109$ $y - 35 = 16$; $y - 35 + 35 = 16 + 35$; $y = 51$

C. $y - 18 = 5$; $y - 18 + 18 = 5 + 18$; $y = 23$ $y - 4.8 = 9.2$; $y - 4.8 + 4.8 = 9.2 + 4.8$; $y = 14$ $y - 5.8 = 15.3$; $y - 5.8 + 5.8 = 15.3 + 5.8$; $y = 21.1$

D. $y - 18.7 = 4.2$; $y - 18.7 + 18.7 = 4.2 + 18.7$; $y = 22.9$ $y - 96 = 107$; $y - 96 + 96 = 107 + 96$; $y = 203$ $y - 62.5 = 83.1$; $y - 62.5 + 62.5 = 83.1 + 62.5$; $y = 145.6$

E. $y - 10.5 = 4.37$; $y - 10.5 + 10.5 = 4.37 + 10.5$; $y = 14.87$ $y - 73 = 196$; $y - 73 + 73 = 196 + 73$; $y = 269$ $y - 275 = 489$; $y - 275 + 275 = 489 + 275$; $y = 764$

F. $y - 14.6 = 9.8$; $y - 14.6 + 14.6 = 9.8 + 14.6$; $y = 24.4$ $y - 8.2 = 100$; $y - 8.2 + 8.2 = 100 + 8.2$; $y = 108.2$ $y - 96 = 1.4$; $y - 96 + 96 = 1.4 + 96$; $y = 97.4$

Page 59

Name________________ Solving equations

Addition and Subtraction Equations

Solve the equations.

A. $n + 23 = 50$; $n = 27$ $x - 16 = 90$; $x = 106$

B. $a + 47 = 85$; $a = 38$ $y + 2.9 = 9.2$; $y = 6.3$ $c + 0.76 = 1.54$; $c = 0.78$

C. $n + 67 = 282$; $n = 215$ $y + 6.9 = 14.5$; $y = 7.6$ $x - 0.58 = 1.39$; $x = 1.97$

D. $y - 77 = 229$; $y = 306$ $c - 167 = 85$; $c = 252$ $n - 25.8 = 19.7$; $n = 45.5$

E. $c - 376 = 488$; $c = 864$ $x - 8.9 = 17.6$; $x = 26.5$ $n + 89 = 134$; $n = 45$

F. $n + 38 = 84$; $n = 46$ $x + 5.6 = 9.2$; $x = 3.6$ $z - 6.85 = 4.76$; $z = 11.61$

Page 60

Answer Key

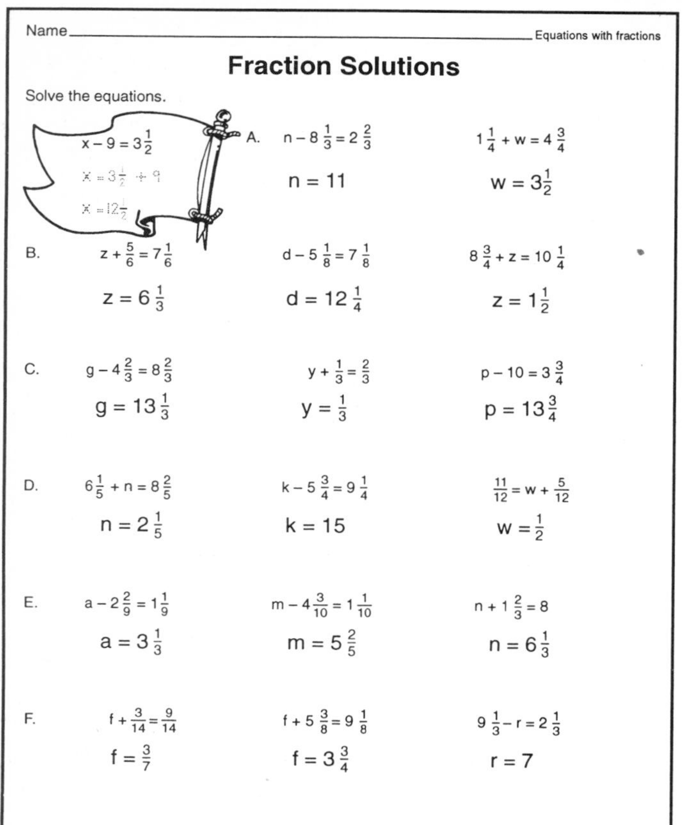

Name________________________ Equations with fractions

Fraction Solutions

Solve the equations.

$x - 9 = 3\frac{1}{2}$
$x = 3\frac{1}{2} + 9$
$x = 12\frac{1}{2}$

A. $n - 8\frac{1}{3} = 2\frac{2}{3}$ $\quad n = 11$ | $1\frac{1}{4} + w = 4\frac{3}{4}$ $\quad w = 3\frac{1}{2}$

B. $z + \frac{5}{6} = 7\frac{1}{6}$ $\quad z = 6\frac{1}{3}$ | $d - 5\frac{1}{8} = 7\frac{1}{8}$ $\quad d = 12\frac{1}{4}$ | $8\frac{3}{4} + z = 10\frac{1}{4}$ $\quad z = 1\frac{1}{2}$

C. $g - 4\frac{2}{3} = 8\frac{2}{3}$ $\quad g = 13\frac{1}{3}$ | $y + \frac{1}{3} = \frac{2}{3}$ $\quad y = \frac{1}{3}$ | $p - 10 = 3\frac{3}{4}$ $\quad p = 13\frac{3}{4}$

D. $6\frac{1}{5} + n = 8\frac{2}{5}$ $\quad n = 2\frac{1}{5}$ | $k - 5\frac{3}{4} = 9\frac{1}{4}$ $\quad k = 15$ | $\frac{11}{12} = w + \frac{5}{12}$ $\quad w = \frac{1}{2}$

E. $a - 2\frac{2}{9} = 1\frac{1}{9}$ $\quad a = 3\frac{1}{3}$ | $m - 4\frac{3}{10} = 1\frac{1}{10}$ $\quad m = 5\frac{2}{5}$ | $n + 1\frac{2}{3} = 8$ $\quad n = 6\frac{1}{3}$

F. $f + \frac{3}{14} = \frac{9}{14}$ $\quad f = \frac{3}{7}$ | $f + 5\frac{3}{8} = 9\frac{1}{8}$ $\quad f = 3\frac{3}{4}$ | $9\frac{1}{3} - r = 2\frac{1}{3}$ $\quad r = 7$

Page 61

Name________________________ Writing algebraic expressions

Say It With Symbols

Translate each expression by writing an equation with numbers and a variable.

The product of 8 and n is 56. $8 \cdot n = 56$

A. The product of 12 and n is 132. $12 \cdot n = 132$

B. 60 divided by i is 15. $60 \div i = 15$ | 48 shared equally among 4 is z. $48 \div 4 = z$

C. F divided by 5 is 5. $f \div 5 = 5$ | A number divided by 8 is 2. $n \div 8 = 2$

D. Double a number r is 12. $2 \cdot r = 12$ | 35 is the product of m and 7. $35 = m \cdot 7$

E. The product of n and 8 is 108. $n \cdot 8 = 108$ | G divided by 25 is 8. $g \div 25 = 8$

F. 82 shared equally among k is 20.5. $82 \div k = 20.5$ | Triple a number d is 96. $3 \cdot d = 96$

G. The product of y and 7 is 84. $y \cdot 7 = 84$ | 19 is the quotient of 76 divided by x. $19 = 76 \div x$

H. 150 divided by w is 6. $150 \div w = 6$ | B divided by 5 is 14. $b \div 5 = 14$

I. 96 divided by n is 4. $96 \div n = 4$ | 8 times v is 168. $8 \cdot v = 168$

J. 46 is the quotient of p divided by 3. $46 = p \div 3$ | 78 is the product of 13 and q. $78 = 13 \cdot q$

Page 62

Name________________________ Function tables

More Number Patterns

Use the rules given to complete the function tables.

A.

x	5	6	7	8	9
3x	15	18	21	24	27

y	10	12	14	16	18
$\frac{y}{2}$	5	6	7	8	9

B.

z	5	10	15	20	25
6z	30	60	90	120	150

b	4	8	12	16	20
$\frac{b}{4}$	1	2	3	4	5

C.

c	20	21	22	23	24
20c	400	420	440	460	480

f	10	20	30	40	50
$\frac{f}{5}$	2	4	6	8	10

D.

h	10	15	20	25	30
$\frac{h}{10}$	1	$1\frac{1}{2}$	2	$2\frac{1}{2}$	3

m	1	2	3	4	5
1.5m	1.5	3	4.5	6	7.5

For each function table, write the rule and complete the table.

E.

g	1	2	3	4	5
3g	3	6	9	12	15

i	5	10	15	20	25
$\frac{i}{5}$	1	2	3	4	5

F.

k	10	12	14	16	18
$\frac{k}{2}$	5	6	7	8	9

n	50	60	70	80	90
$\frac{n}{5}$	10	12	14	16	18

G.

r	3	5	7	9	11
20r	60	100	140	180	220

p	38	57	76	95	114
$\frac{p}{19}$	2	3	4	5	6

Page 63

Name________________________ Evaluating expressions

Multiplication and Division Expressions

Evaluate each expression for the given values.

Let x = 15.

A. $3x = 45$ | $\frac{x}{5} = 3$ | $2x \div 3 = 10$

B. $2.5x = 37.5$ | $105 \div x = 7$ | $\frac{2}{3}x = 10$

Let y = 150. Let z = 32.

C. $\frac{z}{16} = 2$ | $\frac{y}{10} = 15$ | $\frac{2y}{3} = 100$

D. $4y = 600$ | $y \div 30 = 5$ | $\frac{z}{0.5} = 64$

Let a = 5. Let b = 15.

E. $3b \div 5 = 9$ | $\frac{18a}{3} = 30$ | $4b \cdot a = 300$

F. $ab \div 3 = 25$ | $\frac{1}{3}b \div a = 1$ | $\frac{b}{a} = 3$

G. $\frac{ab}{10} = 7\frac{1}{2}$ | $\frac{30}{2a} = 3$ | $\frac{30}{2b} = 1$

Let r = 200. Let s = 28. Let t = 10.

H. $\frac{r}{4} = 50$ | $3s = 84$ | $\frac{10s}{2} = 140$

I. $\frac{r}{2t} = 10$ | $2s \cdot t = 560$ | $\frac{t}{r} = \frac{1}{20}$

J. $\frac{st}{2.8} = 100$ | $9t = 90$ | $\frac{s}{4} \cdot t = 70$

Page 64

Answer Key

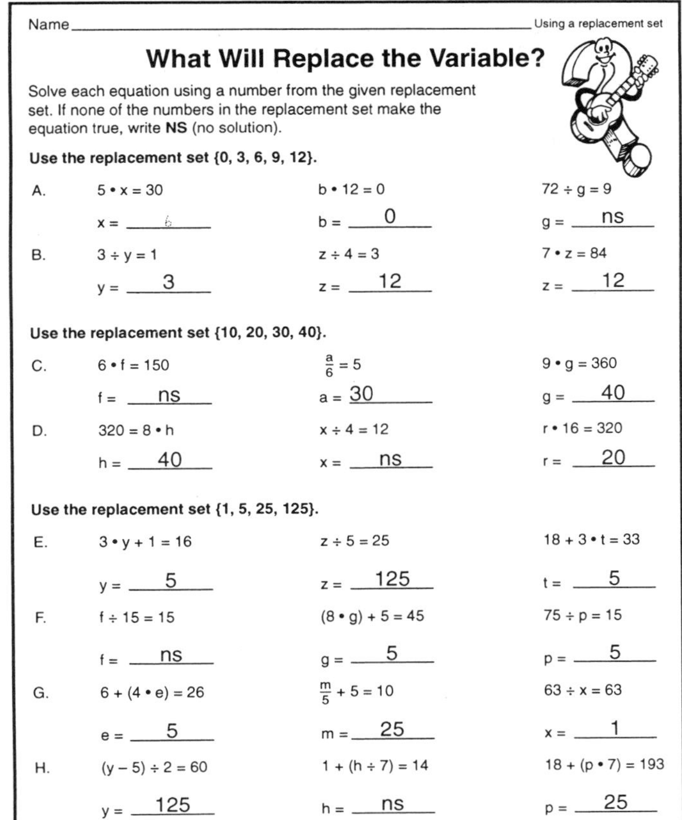

Name ______ Using a replacement set

What Will Replace the Variable?

Solve each equation using a number from the given replacement set. If none of the numbers in the replacement set make the equation true, write **NS** (no solution).

Use the replacement set {0, 3, 6, 9, 12}.

A.	$5 \cdot x = 30$	$b \cdot 12 = 0$	$72 \div g = 9$
	x = 6	b = 0	g = ns
B.	$3 \div y = 1$	$z \div 4 = 3$	$7 \cdot z = 84$
	y = 3	z = 12	z = 12

Use the replacement set {10, 20, 30, 40}.

C.	$6 \cdot f = 150$	$\frac{a}{6} = 5$	$9 \cdot g = 360$
	f = ns	a = 30	g = 40
D.	$320 = 8 \cdot h$	$x \div 4 = 12$	$r \cdot 16 = 320$
	h = 40	x = ns	r = 20

Use the replacement set {1, 5, 25, 125}.

E.	$3 \cdot y + 1 = 16$	$z \div 5 = 25$	$18 + 3 \cdot t = 33$
	y = 5	z = 125	t = 5
F.	$f \div 15 = 15$	$(8 \cdot g) + 5 = 45$	$75 \div p = 15$
	f = ns	g = 5	p = 5
G.	$6 + (4 \cdot e) = 26$	$\frac{m}{5} + 5 = 10$	$63 \div x = 63$
	e = 5	m = 25	x = 1
H.	$(y - 5) \div 2 = 60$	$1 + (h \div 7) = 14$	$18 + (p \cdot 7) = 193$
	y = 125	h = ns	p = 25

Page 65

Name ______ Solving multiplication equations

Multiplication Equations

To solve a multiplication equation, divide both sides by the same number to make the variable stand alone.

$15x = 45$
$15x \div \mathbf{15} = 45 \div \mathbf{15}$
$x = 3$

A.	$6x = 42$	$7x = 112$	$8x = 72$	$246 = 6x$
	x = 7	x = 16	x = 9	x = 41
B.	$7x = 49$	$33x = 66$	$6x = 96$	$13x = 169$
	x = 7	x = 2	x = 16	x = 13
C.	$5x = 280$	$192 = 6x$	$7x = 105$	$144 = 24x$
	x = 56	x = 32	x = 15	x = 6
D.	$13x = 52$	$5.2x = 26$	$21.6 = 5.4x$	$9 = 2.25x$
	x = 4	x = 5	x = 4	x = 4
E.	$248 = 8x$	$15x = 240$	$4.8x = 36$	$56 = 3.5x$
	x = 31	x = 16	x = 7.5	x = 16
F.	$250 = 10x$	$93x = 186$	$1.2x = 14.4$	$200 = 2.5x$
	x = 25	x = 2	x = 12	x = 80

Page 66

Name ______ Solving division equations

$\frac{n}{2} = 14$
$\frac{n}{2} \cdot 2 = 14 \cdot 2$
$n = 28$

Division Equations

To solve a division equation, multiply both sides by the same number to make the variable stand alone.

A.	$\frac{n}{7} = 21$	$\frac{n}{3} = 15$	$\frac{n}{4} = 10$	$\frac{n}{2} = 17$
	n = 147	n = 45	n = 40	n = 34
B.	$\frac{n}{6} = 72$	$\frac{n}{4} = 19$	$\frac{n}{5} = 25$	$\frac{n}{15} = 5$
	n = 432	n = 76	n = 125	n = 75
C.	$\frac{n}{8} = 25$	$\frac{n}{3} = 7.5$	$\frac{n}{15} = 14$	$\frac{n}{8} = 5.9$
	n = 200	n = 22.5	n = 210	n = 47.2
D.	$\frac{n}{3.5} = 12$	$\frac{n}{14} = 2.5$	$\frac{n}{10} = 8\frac{1}{2}$	$\frac{n}{4.7} = 93$
	n = 42	n = 35	n = 85	n = 437.1
E.	$\frac{n}{5} = 4.6$	$\frac{n}{2} = 3\frac{1}{3}$	$\frac{n}{9} = 27$	$\frac{n}{5.1} = 20.4$
	n = 23	$n = 6\frac{2}{3}$	n = 243	n = 104.04
F.	$\frac{n}{10.9} = 2$	$\frac{n}{15.2} = 4.9$	$\frac{n}{6} = 42.7$	$\frac{n}{100} = \frac{2}{5}$
	n = 21.8	n = 74.48	n = 256.2	n = 40

Page 67

Name ______ Solving equations

Multiplication and Division Equations

Solve the equations.

A.	$8b = 72$	$240 = 8x$	$3 = \frac{p}{70}$	
	b = 9	x = 30	p = 210	
B.	$4a = 28$	$10 = \frac{n}{3}$	$350w = 700$	$100 = 5z$
	a = 7	n = 30	w = 2	z = 20
C.	$\frac{x}{8} = 9$	$125e = 250$	$\frac{b}{10} = 8$	$250 = 2.5z$
	x = 72	e = 2	b = 80	z = 100
D.	$\frac{r}{1,000} = 7$	$24b = 312$	$\frac{n}{1.2} = 1.8$	$47n = 423$
	r = 7,000	b = 13	n = 2.16	n = 9
E.	$\frac{t}{27} = 36$	$5.27 = 3.1n$	$23 = \frac{c}{35}$	$\frac{b}{7.5} = 2.8$
	t = 972	n = 1.7	c = 805	b = 21
F.	$\frac{x}{2.5} = 0.25$	$\frac{d}{54} = 1.83$	$0.4c = 68$	$0.09n = 27$
	x = 0.625	d = 98.82	c = 170	n = 300

Page 68

Answer Key

Name______________________ Understanding integers

Understanding Integers

The set of integers contains all positive whole numbers and their negative opposites. Write an integer suggested by each situation listed below.

A.	a savings of $10	+10	a loss of 7 points	−7
B.	a gain of 4 yards	+4	5 miles below sea level	−5
C.	a decrease of 15 pounds	−15	10 seconds before liftoff	−10
D.	3 feet under water	−3	100 feet above sea level	+100
E.	a 12-foot-deep crater	−12	a 15° drop in temperature	−15
F.	an expense of $39	−39	a 20-yard penalty	−20
G.	50 years ago	−50	earnings of $45	+45
H.	a profit of $150	+150	14 years from now	+14
I.	a debt of $175	−175	a stock price drop of $1	−1
J.	a 17° rise in temperature	+17	6 laps behind the lead car	−6
K.	a $25 profit	+25	a $50 bonus	+50

Page 69

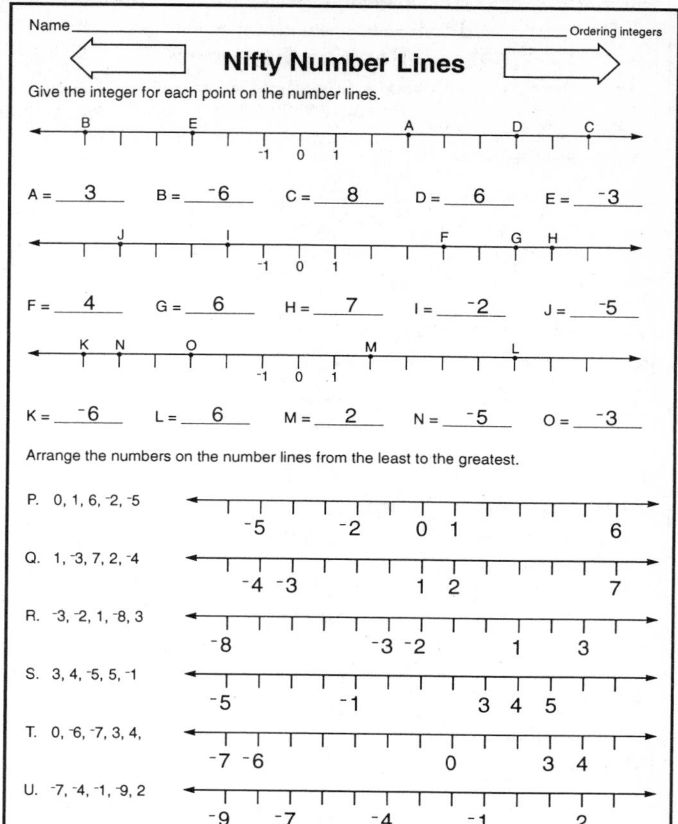

Name______________________ Ordering integers

Nifty Number Lines

Give the integer for each point on the number lines.

A = 3 B = -6 C = 8 D = 6 E = -3

F = 4 G = 6 H = 7 I = -2 J = -5

K = -6 L = 6 M = 2 N = -5 O = -3

Arrange the numbers on the number lines from the least to the greatest.

P. 0, 1, 6, -2, -5 — -5, -2, 0, 1, 6

Q. 1, -3, 7, 2, -4 — -4, -3, 1, 2, 7

R. -3, -2, 1, -8, 3 — -8, -3, -2, 1, 3

S. 3, 4, -5, 5, -1 — -5, -1, 3, 4, 5

T. 0, -6, -7, 3, 4, — -7, -6, 0, 3, 4

U. -7, -4, -1, -9, 2 — -9, -7, -4, -1, 2

Page 70

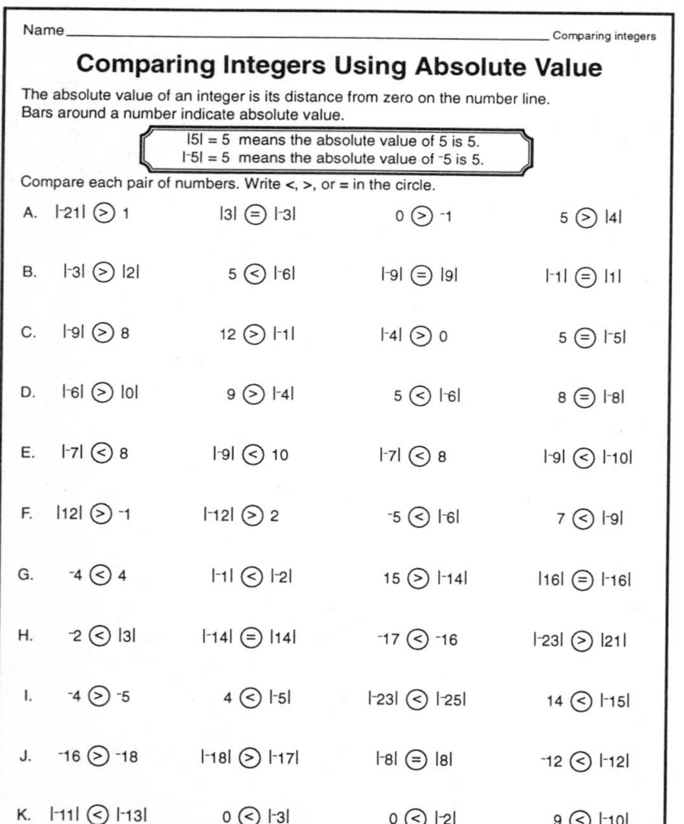

Name______________________ Comparing integers

Comparing Integers Using Absolute Value

The absolute value of an integer is its distance from zero on the number line. Bars around a number indicate absolute value.

|5| = 5 means the absolute value of 5 is 5.
|-5| = 5 means the absolute value of -5 is 5.

Compare each pair of numbers. Write <, >, or = in the circle.

A. |-21| > 1 — |3| = |-3| — 0 > -1 — 5 > |4|

B. |-3| > |2| — 5 < |-6| — |-9| = |9| — |-1| = |1|

C. |-9| > 8 — 12 > |-1| — |-4| > 0 — 5 = |-5|

D. |-6| > |0| — 9 > |-4| — 5 < |-6| — 8 = |-8|

E. |-7| < 8 — |-9| < 10 — |-7| < 8 — |-9| < |-10|

F. |12| > -1 — |-12| > 2 — -5 < |-6| — 7 < |-9|

G. -4 < 4 — |-1| < |-2| — 15 > |-14| — |16| = |-16|

H. -2 < |3| — |-14| = |14| — -17 < -16 — |-23| > |21|

I. -4 > -5 — 4 < |-5| — |-23| < |-25| — 14 < |-15|

J. -16 > -18 — |-18| > |-17| — |-8| = |8| — -12 < |-12|

K. |-11| < |-13| — 0 < |-3| — 0 < |-2| — 9 < |-10|

Page 71

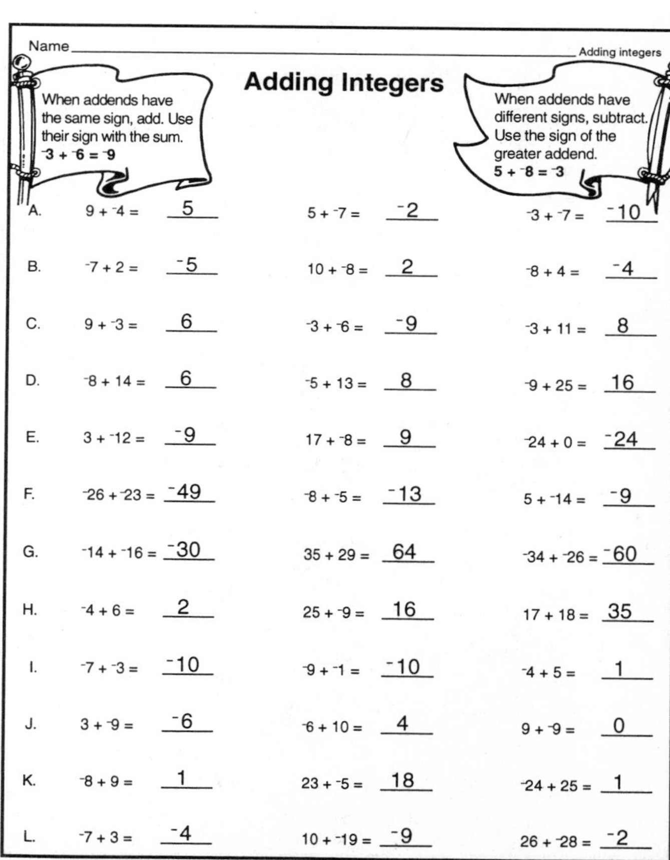

Name______________________ Adding integers

Adding Integers

When addends have the same sign, add. Use their sign with the sum.
-3 + -6 = -9

When addends have different signs, subtract. Use the sign of the greater addend.
5 + -8 = -3

A.	9 + -4 = 5	5 + -7 = -2	-3 + -7 = -10
B.	-7 + 2 = -5	10 + -8 = 2	-8 + 4 = -4
C.	9 + -3 = 6	-3 + -6 = -9	-3 + 11 = 8
D.	-8 + 14 = 6	-5 + 13 = 8	-9 + 25 = 16
E.	3 + -12 = -9	17 + -8 = 9	-24 + 0 = -24
F.	-26 + -23 = -49	-8 + -5 = -13	5 + -14 = -9
G.	-14 + -16 = -30	35 + 29 = 64	-34 + -26 = -60
H.	-4 + 6 = 2	25 + -9 = 16	17 + 18 = 35
I.	-7 + -3 = -10	-9 + -1 = -10	-4 + 5 = 1
J.	3 + -9 = -6	-6 + 10 = 4	9 + -9 = 0
K.	-8 + 9 = 1	23 + -5 = 18	-24 + 25 = 1
L.	-7 + 3 = -4	10 + -19 = -9	26 + -28 = -2

Page 72

Answer Key

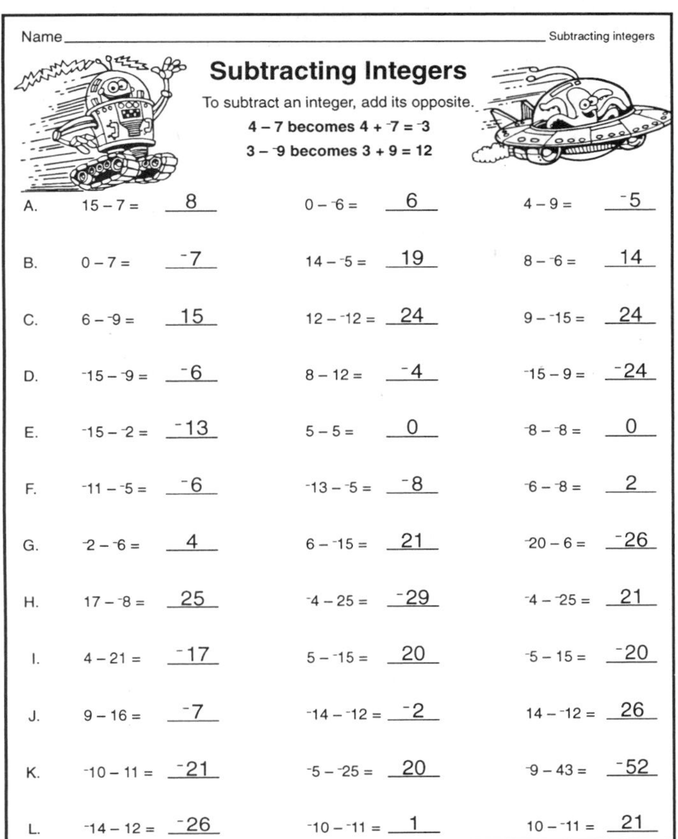

Name________________________ Subtracting integers

Subtracting Integers

To subtract an integer, add its opposite.

$4 - 7$ becomes $4 + {}^{-}7 = {}^{-}3$

$3 - {}^{-}9$ becomes $3 + 9 = 12$

	Problem	Answer	Problem	Answer	Problem	Answer
A.	$15 - 7 =$	8	$0 - {}^{-}6 =$	6	$4 - 9 =$	${}^{-}5$
B.	$0 - 7 =$	${}^{-}7$	$14 - {}^{-}5 =$	19	$8 - {}^{-}6 =$	14
C.	$6 - {}^{-}9 =$	15	$12 - {}^{-}12 =$	24	$9 - {}^{-}15 =$	24
D.	${}^{-}15 - {}^{-}9 =$	${}^{-}6$	$8 - 12 =$	${}^{-}4$	${}^{-}15 - 9 =$	${}^{-}24$
E.	${}^{-}15 - {}^{-}2 =$	${}^{-}13$	$5 - 5 =$	0	${}^{-}8 - {}^{-}8 =$	0
F.	${}^{-}11 - {}^{-}5 =$	${}^{-}6$	${}^{-}13 - {}^{-}5 =$	${}^{-}8$	${}^{-}6 - {}^{-}8 =$	2
G.	${}^{-}2 - {}^{-}6 =$	4	$6 - {}^{-}15 =$	21	${}^{-}20 - 6 =$	${}^{-}26$
H.	$17 - {}^{-}8 =$	25	${}^{-}4 - 25 =$	${}^{-}29$	${}^{-}4 - {}^{-}25 =$	21
I.	$4 - 21 =$	${}^{-}17$	$5 - {}^{-}15 =$	20	${}^{-}5 - 15 =$	${}^{-}20$
J.	$9 - 16 =$	${}^{-}7$	${}^{-}14 - {}^{-}12 =$	${}^{-}2$	$14 - {}^{-}12 =$	26
K.	${}^{-}10 - 11 =$	${}^{-}21$	${}^{-}5 - {}^{-}25 =$	20	${}^{-}9 - 43 =$	${}^{-}52$
L.	${}^{-}14 - 12 =$	${}^{-}26$	${}^{-}10 - {}^{-}11 =$	1	$10 - {}^{-}11 =$	21

Page 73

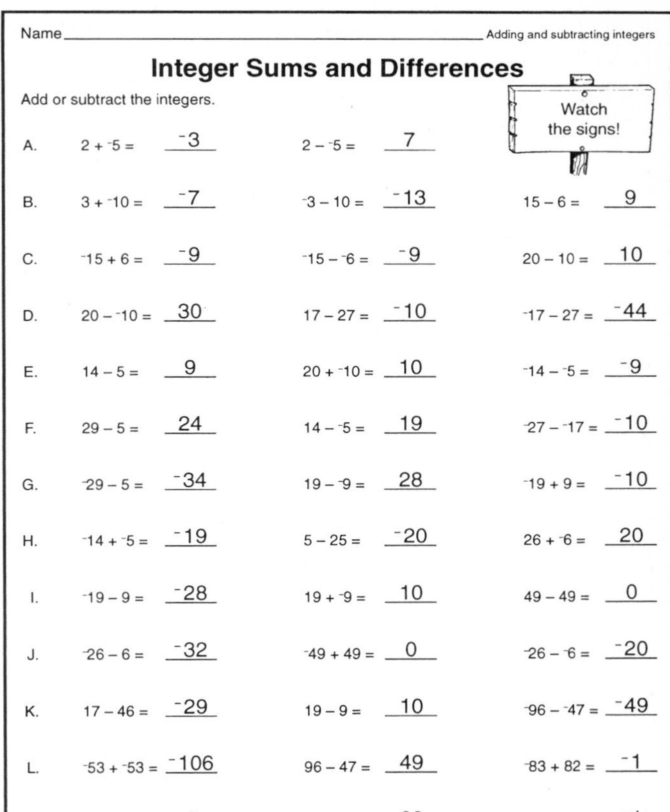

Name________________________ Adding and subtracting integers

Integer Sums and Differences

Add or subtract the integers.

	Problem	Answer	Problem	Answer	Problem	Answer
A.	$2 + {}^{-}5 =$	${}^{-}3$	$2 - {}^{-}5 =$	7		
B.	$3 + {}^{-}10 =$	${}^{-}7$	${}^{-}3 - 10 =$	${}^{-}13$	$15 - 6 =$	9
C.	${}^{-}15 + 6 =$	${}^{-}9$	${}^{-}15 - {}^{-}6 =$	${}^{-}9$	$20 - 10 =$	10
D.	$20 - {}^{-}10 =$	30	$17 - 27 =$	${}^{-}10$	${}^{-}17 - 27 =$	${}^{-}44$
E.	$14 - 5 =$	9	$20 + {}^{-}10 =$	10	${}^{-}14 - {}^{-}5 =$	${}^{-}9$
F.	$29 - 5 =$	24	$14 - {}^{-}5 =$	19	${}^{-}27 - {}^{-}17 =$	${}^{-}10$
G.	${}^{-}29 - 5 =$	${}^{-}34$	$19 - {}^{-}9 =$	28	${}^{-}19 + 9 =$	${}^{-}10$
H.	${}^{-}14 + {}^{-}5 =$	${}^{-}19$	$5 - 25 =$	${}^{-}20$	$26 + {}^{-}6 =$	20
I.	${}^{-}19 - 9 =$	${}^{-}28$	$19 + {}^{-}9 =$	10	$49 - 49 =$	0
J.	${}^{-}26 - 6 =$	${}^{-}32$	${}^{-}49 + 49 =$	0	${}^{-}26 - {}^{-}6 =$	${}^{-}20$
K.	$17 - 46 =$	${}^{-}29$	$19 - 9 =$	10	${}^{-}96 - {}^{-}47 =$	${}^{-}49$
L.	${}^{-}53 + {}^{-}53 =$	${}^{-}106$	$96 - 47 =$	49	${}^{-}83 + 82 =$	${}^{-}1$
M.	${}^{-}53 - {}^{-}53 =$	0	$46 - 17 =$	29	${}^{-}83 - {}^{-}82 =$	${}^{-}1$

Page 74

Name________________________ Multiplying integers

Multiplying Integers

The product is positive if both factors are positive or if both are negative. The product is negative if one factor is positive and one is negative.

	Problem	Answer	Problem	Answer	Problem	Answer
A.	$4 \bullet 5 =$	20	${}^{-}4 \bullet 9 =$	${}^{-}36$	$8 \bullet {}^{-}6 =$	${}^{-}48$
B.	${}^{-}8 \bullet {}^{-}9 =$	72	${}^{-}5 \bullet 7 =$	${}^{-}35$	$2 \bullet {}^{-}3 =$	${}^{-}6$
C.	${}^{-}4 \bullet 11 =$	${}^{-}44$	${}^{-}2 \bullet {}^{-}13 =$	26	$12 \bullet 12 =$	144
D.	${}^{-}9 \bullet {}^{-}7 =$	63	${}^{-}8 \bullet 5 =$	${}^{-}40$	${}^{-}12 \bullet 6 =$	${}^{-}72$
E.	${}^{-}6 \bullet {}^{-}8 =$	48	$29 \bullet {}^{-}6 =$	${}^{-}174$	$5 \bullet 7 =$	35
F.	${}^{-}11 \bullet 10 =$	${}^{-}110$	${}^{-}10 \bullet {}^{-}4 =$	40	$15 \bullet {}^{-}8 =$	${}^{-}120$
G.	${}^{-}9 \bullet 11 =$	${}^{-}99$	$6 \bullet {}^{-}7 =$	${}^{-}42$	${}^{-}4 \bullet {}^{-}6 =$	24
H.	$8 \bullet 7 =$	56	$14 \bullet {}^{-}9 =$	${}^{-}126$	${}^{-}8 \bullet 16 =$	${}^{-}128$
I.	${}^{-}9 \bullet 8 =$	${}^{-}72$	$12 \bullet {}^{-}11 =$	${}^{-}132$	${}^{-}5 \bullet {}^{-}3 =$	15
J.	${}^{-}5 \bullet {}^{-}11 =$	55	${}^{-}8 \bullet 12 =$	${}^{-}96$	${}^{-}2 \bullet {}^{-}7 =$	14
K.	${}^{-}13 \bullet {}^{-}6 =$	78	${}^{-}9 \bullet 5 =$	${}^{-}45$	${}^{-}12 \bullet {}^{-}3 =$	36
L.	$4 \bullet {}^{-}12 =$	${}^{-}48$	$9 \bullet 9 =$	81	${}^{-}11 \bullet {}^{-}11 =$	121
M.	$5 \bullet {}^{-}13 =$	${}^{-}65$	$15 \bullet {}^{-}7 =$	${}^{-}105$	${}^{-}6 \bullet {}^{-}9 =$	54
N.	${}^{-}7 \bullet {}^{-}15 =$	105	${}^{-}14 \bullet 5 =$	${}^{-}70$	$9 \bullet {}^{-}13 =$	${}^{-}117$

Page 75

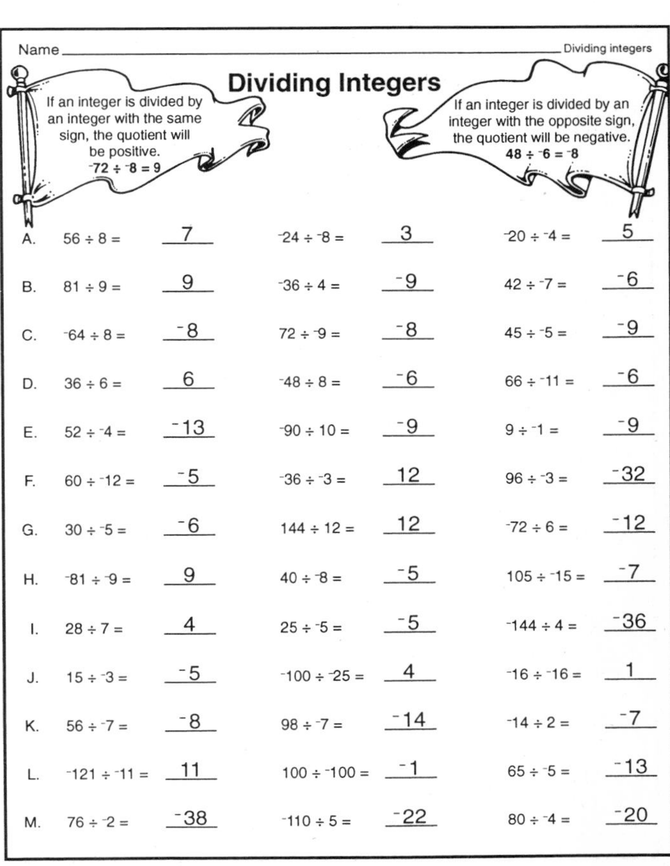

Name________________________ Dividing integers

Dividing Integers

If an integer is divided by an integer with the same sign, the quotient will be positive. **${}^{-}72 \div {}^{-}8 = 9$**

If an integer is divided by an integer with the opposite sign, the quotient will be negative. **$48 \div {}^{-}6 = {}^{-}8$**

	Problem	Answer	Problem	Answer	Problem	Answer
A.	$56 \div 8 =$	7	${}^{-}24 \div {}^{-}8 =$	3	${}^{-}20 \div {}^{-}4 =$	5
B.	$81 \div 9 =$	9	${}^{-}36 \div 4 =$	${}^{-}9$	$42 \div {}^{-}7 =$	${}^{-}6$
C.	${}^{-}64 \div 8 =$	${}^{-}8$	$72 \div {}^{-}9 =$	${}^{-}8$	$45 \div {}^{-}5 =$	${}^{-}9$
D.	$36 \div 6 =$	6	${}^{-}48 \div 8 =$	${}^{-}6$	$66 \div {}^{-}11 =$	${}^{-}6$
E.	$52 \div {}^{-}4 =$	${}^{-}13$	${}^{-}90 \div 10 =$	${}^{-}9$	$9 \div {}^{-}1 =$	${}^{-}9$
F.	$60 \div {}^{-}12 =$	${}^{-}5$	${}^{-}36 \div {}^{-}3 =$	12	$96 \div {}^{-}3 =$	${}^{-}32$
G.	$30 \div {}^{-}5 =$	${}^{-}6$	$144 \div 12 =$	12	${}^{-}72 \div 6 =$	${}^{-}12$
H.	${}^{-}81 \div {}^{-}9 =$	9	$40 \div {}^{-}8 =$	${}^{-}5$	$105 \div {}^{-}15 =$	${}^{-}7$
I.	$28 \div 7 =$	4	$25 \div {}^{-}5 =$	${}^{-}5$	${}^{-}144 \div 4 =$	${}^{-}36$
J.	$15 \div {}^{-}3 =$	${}^{-}5$	${}^{-}100 \div {}^{-}25 =$	4	${}^{-}16 \div {}^{-}16 =$	1
K.	$56 \div {}^{-}7 =$	${}^{-}8$	$98 \div {}^{-}7 =$	${}^{-}14$	${}^{-}14 \div 2 =$	${}^{-}7$
L.	${}^{-}121 \div {}^{-}11 =$	11	$100 \div {}^{-}100 =$	${}^{-}1$	$65 \div {}^{-}5 =$	${}^{-}13$
M.	$76 \div {}^{-}2 =$	${}^{-}38$	${}^{-}110 \div 5 =$	${}^{-}22$	$80 \div {}^{-}4 =$	${}^{-}20$

Page 76

Answer Key

Name ______________________ Multiplying and dividing integers

Integer Products and Quotients

Multiply or divide the integers.

Watch the signs!

A.	5 • -10 =	-50	-16 ÷ 2 =	-8		
B.	-7 • -9 =	63	-12 ÷ 3 =	-4	-12 • -9 =	108
C.	-8 ÷ -2 =	4	-15 ÷ -3 =	5	8 • -8 =	-64
D.	-16 • -6 =	96	-75 ÷ -5 =	15	-48 ÷ 6 =	-8
E.	-80 ÷ 5 =	-16	66 ÷ -11 =	-6	5 • -12 =	-60
F.	-11 • -11 =	121	10 • -9 =	-90	-12 • 5 =	-60
G.	-81 ÷ 9 =	-9	9 • 11 =	99	-90 ÷ 5 =	-18
H.	45 ÷ -9 =	-5	12 • -9 =	-108	-13 • 3 =	-39
I.	-15 • -6 =	90	-12 • -8 =	96	-96 ÷ 8 =	-12
J.	-5 • -8 =	40	-72 ÷ 9 =	-8	-100 ÷ -10 =	10
K.	-11 ÷ 11 =	-1	-8 • -8 =	64	-12 ÷ 1 =	-12
L.	28 ÷ 7 =	4	42 ÷ -6 =	-7	25 • -5 =	-125
M.	12 ÷ -1 =	-12	11 ÷ -11 =	-1	4 • 12 =	48
N.	-12 ÷ -12 =	1	108 ÷ -12 =	-9	11 • -12 =	-132

Page 77

Name ______________________ Integers in expressions

Evaluating Expressions With Integers

Evaluate each expression for the given values.

Let r = -4.

A.	5r =	-20	r + 5 =	1	r – 12 =	-16
B.	$\frac{20}{r}$ =	-5	2.5r =	-10	-6r =	24

Let a = -8. Let b = 6.

C.	a + b =	-2	a – b =	-14	ab =	-48
D.	a – 10 =	-18	-2ab =	96	$\frac{4b}{a}$ =	-3

Let x = -3. Let y = -12.

E.	y + 10 =	-2	-3x =	9	$\frac{y}{3}$ =	-4
F.	y – x =	-9	xy =	36	xy ÷ -4 =	-9
G.	15x =	-45	$\frac{y}{2x}$ =	2	14 + y =	2

Let r = -4. Let s = -1. Let t = -16.

H.	-4r =	16	4rs =	16	t ÷ r =	4
I.	t + 20 =	4	12 – t =	28	15s =	-15
J.	$\frac{t}{4rs}$ =	-1	5r =	-20	$\frac{r}{s}$ =	4

Page 78

Name ______________________ Solving equations with integers

Integer Solutions

Rewrite each equation so that the variable stands alone on one side. Then solve the equation.

Example: x + -7 = -4; x = -4 + 7; x = 3

A.	y – 6 = -18 y = -18 + 6 y = -12	6y = -18 y = $\frac{-18}{6}$ y = -3	t + -9 = 5 t = 5 + 9 t = 14	
B.	-4s = 36 s = $\frac{36}{-4}$ s = -9	-7c = -56 c = $\frac{-56}{-7}$ c = 8	n – -2 = 8 n = 8 – 2 n = 6	y – 7 = -3 y = -3 + 7 y = 4
C.	-5n = -80 n = $\frac{-80}{-5}$ n = 16	x – -14 = 0 x = 0 – 14 x = -14	16 = n – 3 16 + 3 = n n = 19	$\frac{g}{7}$ = -21 g = -21 x 7 g = -147
D.	x – -4 = 5 x = 5 – 4 x = 1	y ÷ -11 = -11 y = -11 x -11 y = 121	72 = -9h $\frac{72}{-9}$ = h h = -8	2t = -64 t = $\frac{-64}{2}$ t = -32
E.	16 + p = 4 p = 4 – 16 p = -12	-30 = 5w $\frac{-30}{5}$ = w w = -6	$\frac{w}{-3}$ = -5 w = -5 x -3 w = 15	-8y = -64 y = $\frac{-64}{-8}$ y = 8
F.	$\frac{c}{9}$ = -2 c = -2 x 9 c = -18	n – -18 = 3 n = 3 – 18 n = -15	z + 8 = -5 z = -5 – 8 z = -13	m ÷ 14 = -3 m = -3 x 14 m = -42

Page 79

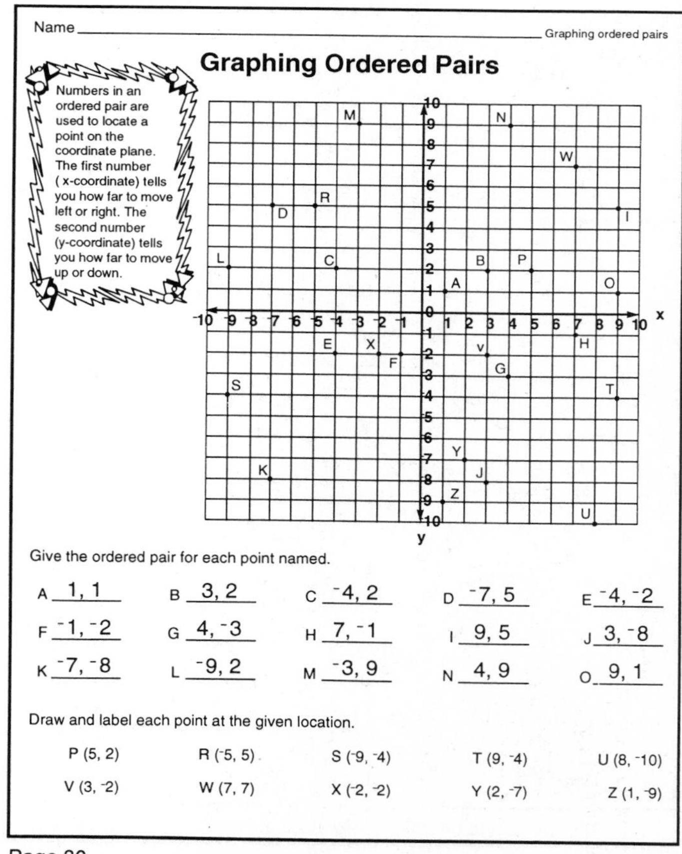
Name ______________________ Graphing ordered pairs

Graphing Ordered Pairs

Numbers in an ordered pair are used to locate a point on the coordinate plane. The first number (x-coordinate) tells you how far to move left or right. The second number (y-coordinate) tells you how far to move up or down.

Give the ordered pair for each point named.

A 1, 1	B 3, 2	C -4, 2	D -7, 5	E -4, -2
F -1, -2	G 4, -3	H 7, -1	I 9, 5	J 3, -8
K -7, -8	L -9, 2	M -3, 9	N 4, 9	O 9, 1

Draw and label each point at the given location.

P (5, 2)	R (-5, 5)	S (-9, -4)	T (9, -4)	U (8, -10)
V (3, -2)	W (7, 7)	X (-2, -2)	Y (2, -7)	Z (1, -9)

Page 80

Answer Key

The Decimal Stops Here

Name ________ Terminating decimals

Change each fraction to a decimal. Your answers will be **terminating decimals** because they end with a remainder of zero. You may use a calculator to check your work.

Remember to keep dividing until the remainder is zero.

A. $\frac{3}{4}$ = 0.75 $\frac{1}{2}$ = 0.5 $\frac{1}{20}$ = 0.05

B. $\frac{5}{2}$ = 2.5 $\frac{17}{50}$ = 0.34 $\frac{4}{5}$ = 0.8

C. $\frac{9}{10}$ = 0.9 $\frac{19}{200}$ = 0.095 $\frac{1}{16}$ = 0.0625 $\frac{19}{40}$ = 0.475

D. $\frac{7}{4}$ = 1.75 $\frac{8}{5}$ = 1.6 $\frac{4}{25}$ = 0.16 $\frac{3}{16}$ = 0.1875

E. $\frac{13}{25}$ = 0.52 $\frac{3}{75}$ = 0.04 $\frac{13}{40}$ = 0.325 $\frac{1}{125}$ = 0.008

F. $\frac{3}{40}$ = 0.075 $\frac{9}{8}$ = 1.125 $\frac{5}{16}$ = 0.3125 $\frac{9}{4}$ = 2.25

G. $\frac{31}{50}$ = 0.62 $\frac{14}{25}$ = 0.56 $\frac{19}{20}$ = 0.95 $\frac{11}{16}$ = 0.6875

H. $\frac{21}{20}$ = 1.05 $\frac{129}{200}$ = 0.645 $\frac{91}{50}$ = 1.82 $\frac{15}{16}$ = 0.9375

I. $\frac{41}{25}$ = 1.64 $\frac{26}{25}$ = 1.04 $\frac{17}{200}$ = 0.085 $\frac{13}{16}$ = 0.8125

J. $\frac{3}{5}$ = 0.6 $\frac{8}{25}$ = 0.32 $\frac{21}{16}$ = 1.3125 $\frac{11}{2}$ = 5.5

K. $\frac{51}{50}$ = 1.02 $\frac{127}{200}$ = 0.635 $\frac{19}{25}$ = 0.76 $\frac{11}{5}$ = 2.2

Page 81

Repeat After Me...

Name ________ Repeating decimals

Repeating decimals do not terminate. You can draw a bar over the digit or digits that repeat. Rewrite each repeating decimal with a bar to show the repeating digit or digits.

0.1444 means the same as $0.1\overline{4}$.

A. 0.16666 = $0.1\overline{6}$ 0.3333 = $0.\overline{3}$

B. 0.161616 = $0.\overline{16}$ 0.263263263 = $0.\overline{263}$

C. 0.161161161 = $0.\overline{161}$ 0.113113113 = $0.\overline{113}$

D. 0.83333 = $0.8\overline{3}$ 0.416666 = $0.41\overline{6}$

Rewrite each fraction as a decimal. Use a calculator to help you. Show the repeating digit or digits with a bar.

E. $\frac{7}{9}$ = $0.\overline{7}$ $\frac{31}{15}$ = $2.0\overline{6}$ $\frac{7}{3}$ = $2.\overline{3}$ $\frac{1}{12}$ = $0.08\overline{3}$

F. $\frac{2}{15}$ = $0.1\overline{3}$ $\frac{19}{12}$ = $1.58\overline{3}$ $\frac{11}{15}$ = $0.7\overline{3}$ $\frac{8}{3}$ = $2.\overline{6}$

G. $\frac{10}{11}$ = $0.\overline{90}$ $\frac{11}{6}$ = $1.8\overline{3}$ $\frac{5}{18}$ = $0.2\overline{7}$ $\frac{11}{12}$ = $0.91\overline{6}$

H. $\frac{7}{60}$ = $0.11\overline{6}$ $\frac{26}{9}$ = $2.\overline{8}$ $\frac{20}{9}$ = $2.\overline{2}$ $\frac{3}{11}$ = $0.\overline{27}$

I. $\frac{22}{3}$ = $7.\overline{3}$ $\frac{13}{36}$ = $0.36\overline{1}$ $\frac{1}{60}$ = $0.01\overline{6}$ $\frac{41}{3}$ = $13.\overline{6}$

J. $\frac{7}{6}$ = $1.1\overline{6}$ $\frac{4}{3}$ = $1.\overline{3}$ $\frac{4}{9}$ = $0.\overline{4}$ $\frac{1}{30}$ = $0.0\overline{3}$

K. $\frac{29}{3}$ = $9.\overline{6}$ $\frac{5}{9}$ = $0.\overline{5}$ $\frac{10}{3}$ = $3.\overline{3}$ $\frac{10}{9}$ = $1.\overline{1}$

Page 82

Decimal Definitions

Name ________ Classifying decimals

Use a calculator to change each fraction to a decimal. Circle **T** or **R** to identify each decimal as **terminating** or **repeating**.

A. $\frac{23}{40}$ = 0.575 (T) R $\frac{11}{3}$ = $3.\overline{6}$ T (R)

B. $\frac{9}{24}$ = 0.375 (T) R $\frac{6}{11}$ = $0.\overline{54}$ T (R)

C. $\frac{16}{25}$ = 0.64 (T) R $\frac{48}{11}$ = $4.\overline{36}$ T (R)

D. $\frac{88}{33}$ = $2.\overline{6}$ T (R) $\frac{72}{99}$ = $0.\overline{72}$ T (R)

E. $\frac{65}{18}$ = $3.6\overline{1}$ T (R) $\frac{44}{54}$ = $0.8\overline{14}$ T (R)

F. $\frac{17}{16}$ = 1.0625 (T) R $\frac{13}{11}$ = $1.\overline{18}$ T (R)

G. $\frac{3}{11}$ = $0.\overline{27}$ T (R) $\frac{47}{50}$ = 0.94 (T) R

H. $\frac{2}{9}$ = $0.\overline{2}$ T (R) $\frac{1}{6}$ = $0.1\overline{6}$ T (R)

I. $\frac{15}{33}$ = $0.\overline{45}$ T (R) $\frac{7}{8}$ = 0.875 (T) R

J. $\frac{33}{16}$ = 2.0625 (T) R $\frac{17}{18}$ = $0.9\overline{4}$ T (R)

K. $\frac{49}{80}$ = 0.6125 (T) R

L. $\frac{8}{9}$ = $0.\overline{8}$ T (R)

Page 83

It's the Greatest!

Name ________ Comparing rational numbers

Compare each pair of numbers. Write $<$, $>$, or $=$. You may want to use a number line.

YOU'RE THE GREATEST

A. $1.5 = 1\frac{1}{2}$ $5\frac{3}{4} < 5.8$

B. $\frac{4}{5} > 0.45$ $2 > \frac{3}{2}$ $\frac{9}{16} < \frac{7}{12}$

C. $3\frac{7}{20} > 3.14$ $0.25 > \frac{1}{5}$ $3\frac{5}{8} > 3.6$

D. $5\frac{3}{25} < 5.15$ $\frac{11}{50} < \frac{6}{25}$ $3\frac{4}{25} = 3.16$

E. $8\frac{3}{5} = 8.6$ $\frac{3}{16} > \frac{1}{6}$ $\frac{1}{9} > 0.11$

F. $4\frac{9}{20} < 4.5$ $4.6 > \frac{35}{8}$ $9\frac{2}{3} > 9.6$

G. $5\frac{1}{8} < 5.2$ $8.3 < 8\frac{1}{3}$ $2.75 = 2\frac{3}{4}$

H. $10\frac{1}{2} > 10.12$ $4.9 = 4\frac{9}{10}$ $3.21 < 3\frac{11}{50}$

I. $8\frac{1}{11} > 8\frac{1}{12}$ $7.5 < 7\frac{5}{9}$ $7\frac{3}{11} > 7\frac{2}{10}$

J. $5\frac{1}{12} < 5.09$ $1.325 < 1\frac{3}{8}$ $7\frac{9}{11} > 7.8$

K. $4\frac{1}{5} = 4.2$ $3.87 < 3\frac{7}{8}$ $5\frac{1}{33} > 5.03$

L. $8\frac{8}{11} < 8.8$ $6.7 < 6\frac{6}{7}$ $9\frac{9}{10} < 9\frac{91}{100}$

Page 84

May I Take Your Order?

Name ________ Ordering rational numbers

Write the numbers in order from the least to the greatest.

A. $^-5.8, ^-7, 3\frac{5}{8}, 1\frac{2}{3}$
$^-7, ^-5.8, 1\frac{2}{3}, 3\frac{5}{8}$

$\frac{7}{9}, \frac{^-5}{12}, \frac{^-7}{36}, ^-0.5, 1.1$
$^-0.5, \frac{^-5}{12}, \frac{^-7}{36}, \frac{7}{9}, 1.1$

B. $^-5, 10, 0.8, \frac{^-9}{10}, ^-12$
$^-12, ^-5, \frac{^-9}{10}, 0.8, 10$

$12\frac{3}{4}, ^-16, 0, \frac{15}{16}, \frac{^-3}{4}$
$^-16, \frac{^-3}{4}, 0, \frac{15}{16}, 12\frac{3}{4}$

C. $^-2\frac{1}{2}, 1, ^-2, 1\frac{1}{2}, 0.5$
$^-2\frac{1}{2}, ^-2, 0.5, 1, 1\frac{1}{2}$

$\frac{3}{5}, 0.45, \frac{^-5}{10}, 0.4$
$\frac{^-5}{10}, 0.4, 0.45, \frac{3}{5}$

D. $1\frac{1}{2}, 2\frac{1}{2}, 1.4, ^-1.55, 2\frac{1}{4}$
$^-1.55, 1.4, 1\frac{1}{2}, 2\frac{1}{4}, 2\frac{1}{2}$

$\frac{^-1}{3}, ^-0.6, 0.3, 0.03, ^-0.03$
$^-0.6, \frac{^-1}{3}, ^-0.03, 0.03, 0.3$

E. $1.45, 1\frac{2}{5}, ^-2.7, ^-2\frac{3}{4}, 1$
$^-2\frac{3}{4}, ^-2.7, 1, 1\frac{2}{5}, 1.45$

$\frac{^-3}{2}, \frac{^-4}{3}, ^-0.5, 0.001, 0$
$\frac{^-3}{2}, \frac{^-4}{3}, ^-0.5, 0, 0.001$

F. $^-1.75, 0.3, 1.5, \frac{^-1}{4}, \frac{^-8}{10}$
$^-1.75, \frac{^-8}{10}, \frac{^-1}{4}, 0.3, 1.5$

$\frac{9}{10}, ^-0.9, ^-1, 0.03, ^-0.05$
$^-1, ^-0.9, ^-0.05, 0.03, \frac{9}{10}$

Page 85

Integer Powers of Ten

Name ________ Integers as exponents

Rewrite each number as a whole number or a fraction. If the exponent is negative, write a fraction with 1 as the numerator and the number with the exponent in positive form as the denominator. Then find the value of the denominator. You may use a calculator.

$5^5 = 3{,}125$ $8^{-2} = \frac{1}{8^2} = \frac{1}{64}$

A. $3^{-4} = \frac{1}{3^4} = \frac{1}{81}$ $2^{-6} = \frac{1}{2^6} = \frac{1}{64}$ $5^4 = 625$

B. $2^{-4} = \frac{1}{2^4} = \frac{1}{16}$ $9^2 = 81$ $4^{-3} = \frac{1}{4^3} = \frac{1}{64}$

C. $7^{-3} = \frac{1}{7^3} = \frac{1}{343}$ $10^{-4} = \frac{1}{10^4} = \frac{1}{10{,}000}$ $8^{-3} = \frac{1}{8^3} = \frac{1}{512}$

D. $2^6 = 64$ $3^3 = 27$ $4^{-4} = \frac{1}{4^4} = \frac{1}{256}$

E. $7^{-2} = \frac{1}{7^2} = \frac{1}{49}$ $2^{-7} = \frac{1}{2^7} = \frac{1}{128}$ $10^{-2} = \frac{1}{10^2} = \frac{1}{100}$

Write each expression as a number with a positive or negative exponent. If the number is a fraction, it will have a negative exponent.

$343 = 7^3$ $\frac{1}{16} = 2^{-4}$

F. $\frac{1}{81} = 9^{-2}$ $\frac{1}{8} = 2^{-3}$ $\frac{1}{9} = 3^{-2}$

G. $\frac{1}{64} = 8^{-2}$ $\frac{1}{25} = 5^{-2}$ $27 = 3^3$

H. $\frac{1}{100} = 10^{-2}$ $1{,}000 = 10^3$ $\frac{1}{121} = 11^{-2}$

I. $\frac{1}{216} = 6^{-3}$ $125 = 5^3$ $\frac{1}{169} = 13^{-2}$

J. $\frac{1}{10{,}000} = 10^{-4}$ $\frac{1}{343} = 7^{-3}$ $\frac{1}{256} = 4^{-4}$

Page 86

Answer Key

Scientific Notation With Integer Exponents

Name ______ Scientific notation

A number is in scientific notation if it is written as the product of a number from 1 to 9 and a power of 10. Write each number using scientific notation. If the standard form number given is less than 1, the exponent will be negative.

0.0023
2.3×10^{-3}

A.	0.93	0.0004	0.0056	0.091
	9.3×10^{-1}	4×10^{-4}	5.6×10^{-3}	9.1×10^{-2}
B.	0.000045	0.0089	158	0.00306
	4.5×10^{-5}	8.9×10^{-3}	1.58×10^{2}	3.06×10^{-3}
C.	0.07345	0.3	0.896	34,967
	7.345×10^{-2}	3×10^{-1}	8.96×10^{-1}	3.4967×10^{4}

Write the numbers in standard form. Watch the signs on the exponents!

D.	9×10^{-3}	4×10^{-4}	8.3×10^{-2}	1.95×10^{-3}
	.009	0.0004	0.083	0.00195
E.	9.2×10^{3}	2.3×10^{-1}	6.03×10^{-2}	8.97×10^{-4}
	9,200	0.23	0.0603	0.000897
F.	7.25×10^{-4}	8.1×10^{-3}	4.083×10^{4}	6.98×10^{-3}
	0.000725	0.0081	40,830	0.00698
G.	5.5×10^{-5}	4.78×10^{-1}	5.834×10^{3}	1.3×10^{-2}
	0.000055	0.478	5,834	0.013

Page 87

Square Roots

Name ______ Square roots

To find the square root of a number, find the number that when multiplied by itself is equal to the number. Every positive number has a positive square root and a negative square root.

A.	$\sqrt{144} = 12$	$\sqrt{16} = 4$	$\sqrt{49} = 7$	
B.	$-\sqrt{25} = -5$	$\sqrt{\frac{1}{4}} = 0.5$	$\sqrt{169} = 13$	
C.	$\sqrt{64} = 8$	$\sqrt{100} = 10$	$-\sqrt{121} = -11$	$\sqrt{81} = 9$
D.	$\sqrt{\frac{1}{16}} = 0.25$	$-\sqrt{400} = -20$	$\sqrt{\frac{9}{25}} = 0.6$	$\sqrt{121} = 11$
E.	$\sqrt{\frac{4}{81}} = 0.\overline{2}$	$\sqrt{0.04} = 0.2$	$-\sqrt{36} = -6$	$\sqrt{\frac{4}{9}} = 0.\overline{6}$
F.	$\sqrt{0.64} = 0.8$	$\sqrt{0.25} = 0.5$	$-\sqrt{0.16} = -0.4$	$\sqrt{\frac{16}{49}} = 0.5714286$

Use a calculator to find the square root of each number below. Round your answers to the nearest tenth.

G.	$\sqrt{56} = 7.5$	$\sqrt{13} = 3.6$	$\sqrt{91} = 9.5$	$\sqrt{21} = 4.6$
H.	$\sqrt{110} = 10.5$	$\sqrt{87} = 9.3$	$\sqrt{250} = 15.8$	$\sqrt{17} = 4.1$
I.	$\sqrt{46} = 6.8$	$\sqrt{112} = 10.6$	$\sqrt{70} = 8.4$	$\sqrt{19} = 4.4$
J.	$\sqrt{57} = 7.5$	$\sqrt{83} = 9.1$	$\sqrt{7} = 2.6$	$\sqrt{96} = 9.8$
K.	$\sqrt{30} = 5.5$	$\sqrt{58} = 7.6$	$\sqrt{10} = 3.2$	$\sqrt{2} = 1.4$
L.	$\sqrt{53} = 7.3$	$\sqrt{150} = 12.2$	$\sqrt{3} = 1.7$	$\sqrt{24} = 4.9$

Page 88

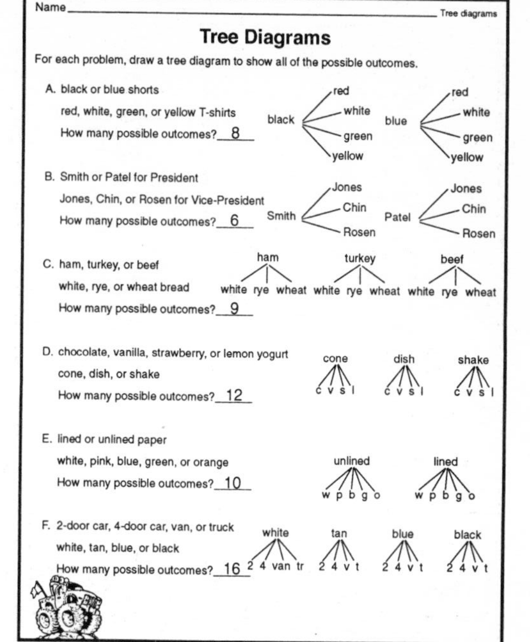

Tree Diagrams

Name ______ Tree diagrams

For each problem, draw a tree diagram to show all of the possible outcomes.

A. black or blue shorts
red, white, green, or yellow T-shirts
How many possible outcomes? 8

B. Smith or Patel for President
Jones, Chin, or Rosen for Vice-President
How many possible outcomes? 6

C. ham, turkey, or beef
white, rye, or wheat bread
How many possible outcomes? 9

D. chocolate, vanilla, strawberry, or lemon yogurt
cone, dish, or shake
How many possible outcomes? 12

E. lined or unlined paper
white, pink, blue, green, or orange
How many possible outcomes? 10

F. 2-door car, 4-door car, van, or truck
white, tan, blue, or black
How many possible outcomes? 16

Page 89

Topping Combinations

Name ______ Exploring combinations

A **combination** is an arrangement of objects or events in which order does not matter.

How many combinations of 2 sundae toppings are possible? First, write down all the nuts combinations, then the chips combinations, then the fudge combinations, and finally the sprinkles combinations. If two lists contain the same combinations, cross out one of the combinations.

Sundae Toppings
nuts chips
fudge sprinkles

nuts, chips	~~chips, nuts~~	~~fudge, nuts~~	~~sprinkles, nuts~~
nuts, fudge	chips, fudge	fudge, sprinkles	~~sprinkles, chips~~
nuts, sprinkles	chips, sprinkles	~~fudge, chips~~	~~sprinkles, fudge~~

3 combinations + 2 combinations + 1 combination + 0 combinations = 6 combinations

Answer the questions about the pizza toppings. Work on scratch paper.

Pizza Toppings
mushrooms
pepperoni
sausage
pineapple
green pepper
olives

A. How many combinations of two toppings are possible? 15

B. How many combinations of two toppings are possible if you don't want sausage or pepperoni? 6

C. How many combinations of two toppings are possible if you don't want pineapple? 10

D. How many combinations of two toppings are possible if you can add extra cheese as a topping? 21

Page 90

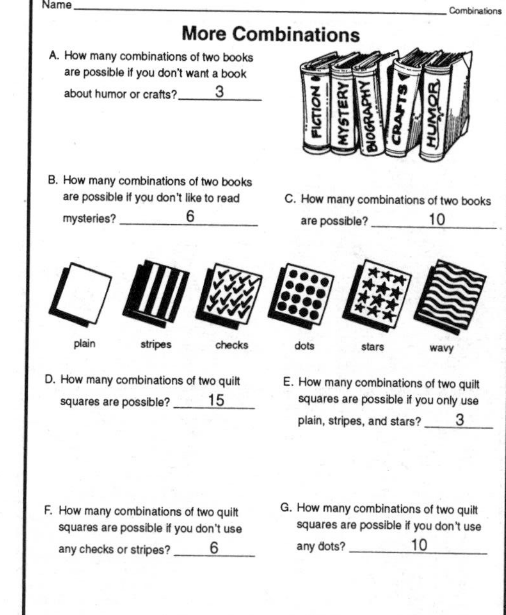

More Combinations

Name ______ Combinations

A. How many combinations of two books are possible if you don't want a book about humor or crafts? 3

B. How many combinations of two books are possible if you don't like to read mysteries? 6

C. How many combinations of two books are possible? 10

D. How many combinations of two quilt squares are possible? 15

E. How many combinations of two quilt squares are possible if you only use plain, stripes, and stars? 3

F. How many combinations of two quilt squares are possible if you don't use any checks or stripes? 6

G. How many combinations of two quilt squares are possible if you don't use any dots? 10

Page 91

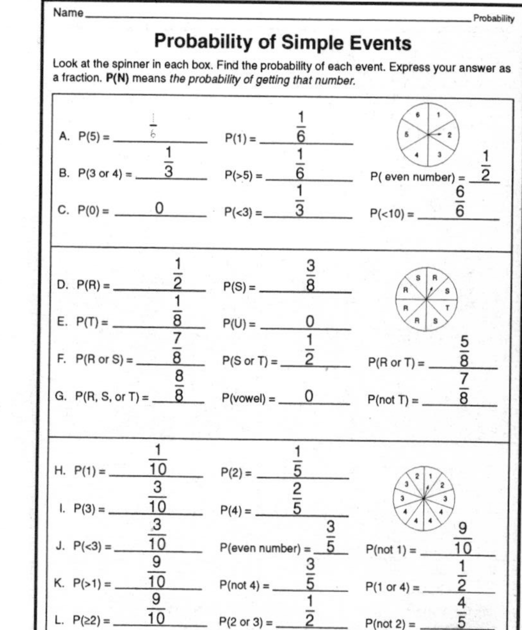

Probability of Simple Events

Name ______ Probability

Look at the spinner in each box. Find the probability of each event. Express your answer as a fraction. **P(N)** means *the probability of getting that number.*

A. P(5) = $\frac{1}{6}$ P(1) = $\frac{1}{6}$
B. P(3 or 4) = $\frac{1}{3}$ P(>5) = $\frac{1}{6}$ P(even number) = $\frac{1}{2}$
C. P(0) = 0 P(<3) = $\frac{1}{3}$ P(<10) = $\frac{6}{6}$

D. P(R) = $\frac{1}{2}$ P(S) = $\frac{3}{8}$
E. P(T) = $\frac{1}{8}$ P(U) = 0
F. P(R or S) = $\frac{7}{8}$ P(S or T) = $\frac{1}{2}$ P(R or T) = $\frac{5}{8}$
G. P(R, S, or T) = $\frac{8}{8}$ P(vowel) = 0 P(not T) = $\frac{7}{8}$

H. P(1) = $\frac{1}{10}$ P(2) = $\frac{1}{5}$
I. P(3) = $\frac{3}{10}$ P(4) = $\frac{2}{5}$
J. P(<3) = $\frac{3}{10}$ P(even number) = $\frac{3}{5}$ P(not 1) = $\frac{9}{10}$
K. P(>1) = $\frac{9}{10}$ P(not 4) = $\frac{3}{5}$ P(1 or 4) = $\frac{1}{2}$
L. P(≥2) = $\frac{9}{10}$ P(2 or 3) = $\frac{1}{2}$ P(not 2) = $\frac{4}{5}$

Page 92

Answer Key

Independent Events

Name ______________________ Probability

Find the probability of the events in each box. Express your answers as fractions.

A. P(1, A) = $\frac{1}{4} \cdot \frac{1}{3} = \frac{1}{12}$ P(2, B) = $\frac{1}{4} \cdot \frac{1}{3} = \frac{1}{12}$

B. P(1, A or B) = $\frac{1}{4} \cdot \frac{2}{3} = \frac{1}{6}$ P(1 or 2, C) = $\frac{1}{2} \cdot \frac{1}{3} = \frac{1}{6}$

C. P(1, D) = $\frac{1}{4} \cdot 0 = 0$ P(even, A) = $\frac{1}{2} \cdot \frac{1}{3} = \frac{1}{6}$ P(2, vowel) = $\frac{1}{4} \cdot \frac{1}{3} = \frac{1}{12}$

D. P(M, A) = $\frac{1}{4} \cdot \frac{1}{5} = \frac{1}{20}$ P(T, O) = $\frac{1}{4} \cdot \frac{1}{5} = \frac{1}{20}$

E. P(M, R) = $\frac{1}{4} \cdot 0 = 0$ P(S, vowel) = $\frac{1}{4} \cdot 1 = \frac{1}{4}$

F. P(T, not A) = $\frac{1}{4} \cdot \frac{4}{5} = \frac{1}{5}$ P(not R, A) = $\frac{3}{4} \cdot \frac{1}{5} = \frac{3}{20}$ P(T, A or E) = $\frac{1}{4} \cdot \frac{2}{5} = \frac{1}{10}$

G. P(T or R, O) = $\frac{1}{2} \cdot \frac{1}{5} = \frac{1}{10}$ P(R or M, not A) = $\frac{1}{2} \cdot \frac{4}{5} = \frac{2}{5}$ P(not S, I or O) = $\frac{3}{4} \cdot \frac{2}{5} = \frac{3}{10}$

H. P(5, 5) = $\frac{1}{6} \cdot \frac{1}{6} = \frac{1}{36}$ P(6, 2) = $\frac{1}{6} \cdot \frac{1}{6} = \frac{1}{36}$

I. P(4, not 2) = $\frac{1}{6} \cdot \frac{5}{6} = \frac{5}{36}$ P(not 1, 6) = $\frac{5}{6} \cdot \frac{1}{6} = \frac{5}{36}$

J. P(3, >4) = $\frac{1}{6} \cdot \frac{1}{3} = \frac{1}{18}$ P(≥4, 1) = $\frac{1}{2} \cdot \frac{1}{6} = \frac{1}{12}$ P(odd, odd) = $\frac{1}{2} \cdot \frac{1}{2} = \frac{1}{4}$

K. P(even, odd) = $\frac{1}{2} \cdot \frac{1}{2} = \frac{1}{4}$ P(even, 3) = $\frac{1}{2} \cdot \frac{1}{6} = \frac{1}{12}$ P(0, 5) = $0 \cdot \frac{1}{6} = 0$

L. P(not 2, not 3) = $\frac{5}{6} \cdot \frac{5}{6} = \frac{25}{36}$ P(not 1, even) = $\frac{5}{6} \cdot \frac{1}{2} = \frac{5}{12}$ P(5, not 3) = $\frac{1}{6} \cdot \frac{5}{6} = \frac{5}{36}$

Page 93

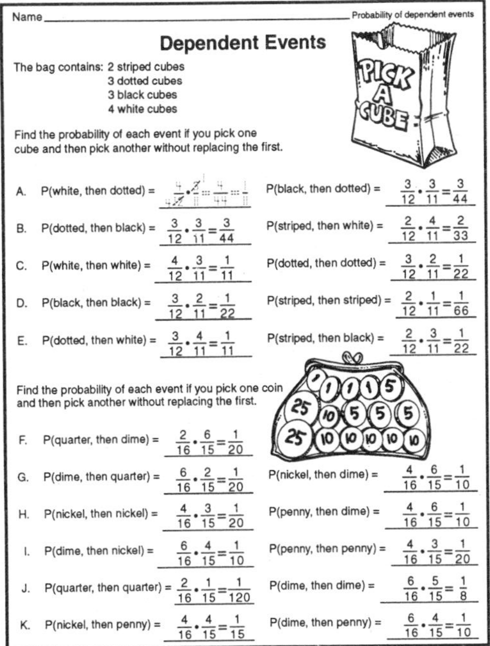

Page 94

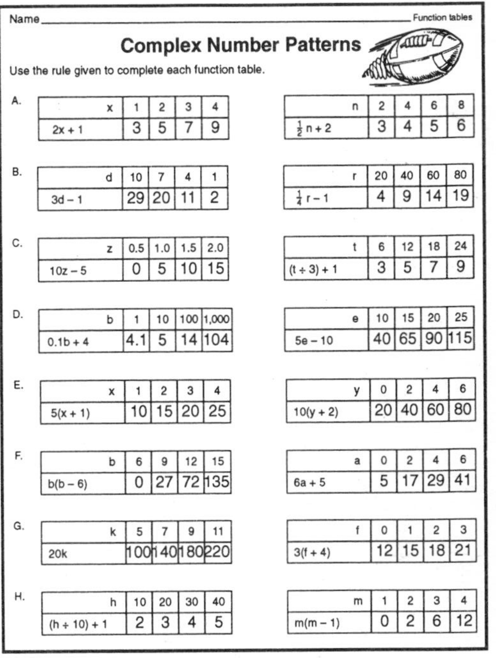

Page 95

Evaluating Complex Expressions

Name ______________________ Evaluating expressions

Evaluate each expression for the given values.

Let x = $^-5$.

A. $x + 2.7 = {}^-2.3$ $12 + x = 7$

B. $x - 4.5 = {}^-9.5$ $5 \cdot x = {}^-25$ $\frac{40}{x} = {}^-8$

Let m = 25.

C. $m + 17.5 = 42.5$ $43 - m = 18$ $m - 19.3 = 5.7$

D. $m^2 = 625$ $3 \cdot m - 20 = 55$ $\frac{m}{10} = 2.5$

Let r = 8. Let s = $^-9$.

E. $(r + s) - 10 = {}^-11$ $r \cdot s = {}^-72$ $\frac{r \cdot s}{2} = {}^-36$

F. $r^2 + s = 55$ $\frac{9r}{s} = {}^-8$ $s - r = {}^-17$

G. $s^2 - r^2 = 17$ $2 \cdot r - 3 = 13$ $r \cdot s - 2 = {}^-74$

Let a = $^-5$. Let b = 8. Let c = $^-9$.

H. $(c - b) + 26 = 9$ $2 \cdot b - 8 = 8$ $(c + a) - 8 = {}^-22$

I. $3.5 + a \cdot b = {}^-36.5$ $a^2 \cdot b = 200$ $\frac{a \cdot b}{10} = {}^-4$

J. $c^2 - b^2 = 17$ $\frac{b \cdot c}{a^2 + 11} = {}^-2$ $(a + b) - 10 = {}^-7$

K. $(8 + b) - c = 25$ $b^2 - a \cdot c = 19$ $b^2 - 32.5 = 31.5$

Page 96

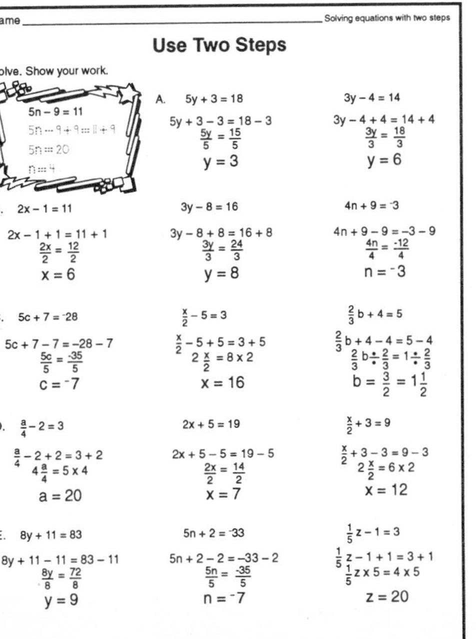

Page 97

Writing and Solving Complex Equations

Name ______________________ Writing and solving equations

Write an equation for each problem. Then solve for the variable.

2 times s decreased by 1 is 9.
$2s - 1 = 9$
$2s = 10$
$s = 5$

A. 4 more than 4 times x is 45.
$4x + 4 = 45$
$4x = 41$
$x = 10.25$

B. 24 less than triple the value of g is 15.
$3g - 24 = 15$
$3g = 39$
$g = 13$

1 more than e divided by 5 is 2.
$\frac{e}{5} + 1 = 2$
$\frac{e}{5} = 1$
$e = 5$

C. 3 more than the product of y and 5 is 38.
$5y + 3 = 38$
$5y = 35$
$y = 7$

6 less than the quotient of h divided by 2 is 10.
$\frac{h}{2} - 6 = 10$
$\frac{h}{2} = 16$
$h = 32$

D. 35 more than twice the value of e is 65.
$2e + 35 = 65$
$2e = 30$
$e = 15$

3 times w increased by 17 is 62.
$3w + 17 = 62$
$3w = 45$
$w = 15$

E. 8 increased by the product of k and 6 is 50.
$6k + 8 = 50$
$6k = 42$
$k = 7$

10 less than the quotient of d divided by 5 is 10.
$\frac{d}{5} - 10 = 10$
$\frac{d}{5} = 20$
$d = 100$

Page 98

Answer Key

Page 99

Name ______ Using the distributive property

Distribute and Solve

Solve the equations.

Example: $4(x+6)=36$
$4x+24=36$
$4x=12$
$x=3$

A.		$3(x+2)=12$ $3x+6=12$ $3x=6$ $x=2$	$4(x-3)=8$ $4x-12=8$ $4x=20$ $x=5$
B.	$2(x+7)=20$ $2x+14=20$ $2x=6$ $x=3$	$5(x-5)=5$ $5x-25=5$ $5x=30$ $x=6$	$6(x-7)=12$ $6x-42=12$ $6x=54$ $x=9$
C.	$5(x-3)=5$ $5x-15=5$ $5x=20$ $x=4$	$7(x-2)=28$ $7x-14=28$ $7x=42$ $x=6$	$5(x+2)=40$ $5x+10=40$ $5x=30$ $x=6$
D.	$2(x+8)=20$ $2x+16=20$ $2x=4$ $x=2$	$6(x+2)=54$ $6x+12=54$ $6x=42$ $x=7$	$3(x-3)=30$ $3x-9=30$ $3x=39$ $x=13$
E.	$4(x+6)=60$ $4x+24=60$ $4x=36$ $x=9$	$8(x-4)=64$ $8x-32=64$ $8x=96$ $x=12$	$7(x+3)=91$ $7x+21=91$ $7x=70$ $x=10$
F.	$9(x-2)=0$ $9x-18=0$ $9x=18$ $x=2$	$5(x+9)=100$ $5x+45=100$ $5x=55$ $x=11$	$8(x-8)=8$ $8x-64=8$ $8x=72$ $x=9$

Page 100

Name ______ Using the distributive property

Distribute and Solve

Solve the equations. Remember to multiply both numbers inside the parentheses by the number in front of the parentheses.

A.	$3(x-2)=-3$ $3x-6=-3$ $3x=3$ $x=1$	$2(x+5)=-2$ $2x+10=-2$ $2x=-12$ $x=-6$	$-4(x-5)=24$ $-4x+20=24$ $-4x=4$ $x=-1$
B.	$-6(x+7)=12$ $-6x-42=12$ $-6x=54$ $x=-9$	$-4(m+12)=36$ $-4m-48=36$ $-4m=84$ $m=-21$	$2(x-4)=22$ $2x-8=22$ $2x=30$ $x=15$
C.	$20=4(x+3)$ $20=4x+12$ $8=4x$ $2=x$	$3(y-7)=27$ $3y-21=27$ $3y=48$ $y=16$	$3(a-5)=-21$ $3a-15=-21$ $3a=-6$ $a=-2$
D.	$4(y+8)=22$ $4y+32=22$ $4y=-10$ $y=-2.5$	$35=5(n+2)$ $35=5n+10$ $25=5n$ $5=n$	$6(z+4)=-10$ $6z+24=-10$ $6z=-34$ $z=-5\frac{2}{3}$
E.	$-5(a+7)=5$ $-5a-35=5$ $-5a=40$ $a=-8$	$3(b-4)=-6$ $3b-12=-6$ $3b=6$ $b=2$	$-18=3(x-4)$ $-18=3x-12$ $-6=3x$ $-2=x$
F.	$-6(x-1)=12$ $-6x+6=12$ $-6x=6$ $x=-1$	$-9(x+1)=18$ $-9x-9=18$ $-9x=27$ $x=-3$	$4(x-5)=-42$ $4x-20=-42$ $4x=-22$ $x=-5.5$

Page 101

Name ______ Combining like terms

Combine the Variables

Multiply the number and variable inside each parentheses by the number in front of the parentheses. Then combine the like variables and solve the equation.

Example: $9x+5(x+7)=-49$
$9x+5x+35=-49$
$14x=-49-35$
$14x=-84$
$x=-6$

A.		$2b+3(b-7)=44$ $2b+3b-21=44$ $5b=65$ $b=13$	$2v+3(8-v)=-16$ $2v+24-3v=-16$ $-v=-40$ $v=40$
B.	$4(2y+9)+7y=-24$ $8y+36+7y=-24$ $15y=-60$ $y=-4$	$5(3n)+14+6n=21$ $15n+14+6n=21$ $21n=7$ $n=\frac{1}{3}$	$5(n+3)+5=-25$ $5n+15+5=-25$ $5n=-45$ $n=-9$
C.	$-8a+6(a+7)=1$ $-8a+6a+42=1$ $-2a=-41$ $a=20\frac{1}{2}$	$10z+5(z-12)=0$ $10z+5z-60=0$ $15z=60$ $z=4$	$7y+7(y+3)=-21$ $7y+7y+21=-21$ $14y=-42$ $y=-3$
D.	$-3x+6(x+4)=9$ $-3x+6x+24=9$ $3x=-15$ $x=-5$	$4z+9+3(2z)=129$ $4z+9+6z=129$ $10z=120$ $z=12$	$6(3a-2)+5a=57$ $18a-12+5a=57$ $23a=69$ $a=3$
E.	$5x+3(x+4)=76$ $5x+3x+12=76$ $8x=64$ $x=8$	$-3(x+4)-21x=72$ $-3x-12-21x=72$ $-24x=84$ $x=-3\frac{1}{2}$	$-4(3+f)+6f=-24$ $-12-4f+6f=-24$ $2f=-12$ $f=-6$

Page 102

Name ______ Solving equations with variables on both sides

On Both Sides Now

Arrange the variables so that they are on the same side of the equation. Then solve the equation.

Example: $5h=12+3h$
$5h-3h=12$
$2h=12$
$h=6$

A.		$2x+72=4x$ $4x-2x=72$ $2x=72$ $x=36$	$24+y=9y$ $9y-y=24$ $8y=24$ $y=3$
B.	$9b=26-4b$ $9b+4b=26$ $13b=26$ $b=2$	$7n=15-8n$ $7n+8n=15$ $15n=15$ $n=1$	$12h=48-4h$ $12h+4h=48$ $16h=48$ $h=3$
C.	$42+9c=16c$ $16c-9c=42$ $7c=42$ $c=6$	$7f=3f-52$ $7f-3f=-52$ $4f=-52$ $f=-13$	$9r+7=4r-8$ $9r+15=4r$ $15=4r-9r$ $15=-5r$ $-3=r$
D.	$-5w-9=3w+7$ $-5w=3w+16$ $-5w-3w=16$ $-8w=16$ $w=-2$	$72+7d=15d$ $72=15d-7d$ $72=8d$ $9=d$	$7r=-5r+144$ $7r+5r=144$ $12r=144$ $r=12$
E.	$-3g+9=15g-9$ $-3g=15g-18$ $-18g=-18$ $g=1$	$5e=14-2e$ $5e+2e=14$ $7e=14$ $e=2$	$3s+10=5s+4$ $3s+6=5s$ $5s-3s=6$ $2s=6$ $s=3$
F.	$6(s+1)=4s+8$ $6s+6=4s+8$ $6s=4s+2$ $2s=2$ $s=1$	$7b+6=2b-19$ $7b+25=2b$ $25=2b-7b$ $25=-5b$ $-5=b$	$8x-1=23-4x$ $8x=24-4x$ $4x+8x=24$ $12x=24$ $x=2$

Page 103

Name ______ Solving equations

Be an Equation Buster!

Solve each equation. Show your work.

Example: $30=8+2x$
$30-8=2x$
$22=2x$
$x=11$

A.		$5a+7=-23$ $5a=-23-7$ $5a=-30$ $a=-6$	$x+38=86-3x$ $x=48-3x$ $4x=48$ $x=12$
B.	$5n=2n+6$ $5n-2n=6$ $3n=6$ $n=2$	$y=24-3y$ $y+3y=24$ $4y=24$ $y=6$	$-12n=35-5n$ $-12n+5n=35$ $-7n=35$ $n=-5$
C.	$\frac{-x}{3}=7$ $-x=21$ $x=-21$	$4n+5=6n+7$ $4n=6n+2$ $-2n=2$ $n=-1$	$2(x-6)=3x$ $2x-12=3x$ $-12=3x-2x$ $-12=x$
D.	$3p-8=13-4p$ $3p=21-4p$ $7p=21$ $p=3$	$8(5-n)=2n$ $40-8n=2n$ $10n=40$ $n=4$	$5(2+n)=3n+18$ $10+5n=3n+18$ $5n=3n+8$ $2n=8$ $n=4$
E.	$4u-8=-5(1-u)$ $4u-8=-5+5u$ $4u=3+5u$ $4u-5u=3$ $-1u=3$ $u=-3$	$2(x-6)=3x$ $2x-12=3x$ $-12=3x-2x$ $-12=x$	$\frac{x}{2}+5=6$ $\frac{x}{2}=1$ $2\frac{x}{2}=1\times 2$ $x=2$
F.	$98-4b=-11b$ $98=-11b+4b$ $98=-7b$ $-14=b$	$\frac{1}{2}(x+8)=10$ $\frac{1}{2}x+4=10$ $\frac{1}{2}x=6$ $x=12$	$3(m+5)=2m+10$ $3m+15=2m+10$ $3m+5=2m$ $5=-1m$ $-5=m$

Page 104

Name ______ Solving equations

More Equation Busting

Solve each equation. Show your work.

Example: $\frac{3}{4}x=12+\frac{1}{4}x$
$\frac{3}{4}x-\frac{1}{4}x=12$
$\frac{2}{4}x=12$
$x=12\div\frac{2}{4}$
$x=24$

A.		$8a=2a+30$ $8a-2a=30$ $6a=30$ $a=5$	$2b=80-8b$ $2b+8b=80$ $10b=80$ $b=8$
B.	$\frac{2}{3}x=6-\frac{1}{3}x$ $\frac{1}{3}x+\frac{1}{3}x=6$ $x=6$	$51=9-3x$ $42=-3x$ $-14=x$	$39c=33c-78$ $39c-33c=-78$ $6c=-78$ $c=-13$
C.	$5p-9=2p+12$ $5p-2p=21$ $3p=21$ $p=7$	$89+x=2-2x$ $87=-3x$ $-29=x$	$4(y-6)=7y$ $4y-24=7y$ $-24=7y-4y$ $-24=3y$ $-8=y$
D.	$7(10-m)=3m$ $70-7m=3m$ $70=10m$ $7=m$	$\frac{2}{3}x-7=x$ $-7=x-\frac{2}{3}x$ $-7=\frac{1}{3}x$ $-21=x$	$3(30+s)=4(s+19)$ $90+3s=4s+76$ $90-76=4s-3s$ $14=s$
E.	$-7a=-12a+65$ $-7a+12a=65$ $5a=65$ $a=13$	$4(x+2)=6x+10$ $4x+8=6x+10$ $-2x=2$ $x=-1$	$\frac{x}{2}+5=x$ $x+10=2x$ $10=2x-x$ $10=x$
F.	$3(2+v)=5v+16$ $6+3v=5v+16$ $-10=2v$ $-5=v$	$4(3y-1)=5y-11$ $12y-4=5y-11$ $7y=-7$ $y=-1$	$5x+2(1-x)=2(2x-1)$ $5x+2-2x=4x-2$ $3x=2=4x-2$ $3x=4x-4$ $x=4$